The Promise of Artificial Intelligence

Reckoning & Judgment

The Promise of Artificial Intelligence

Reckoning and Judgment

Brian Cantwell Smith

The MIT Press
Cambridge, Massachusetts
London, England

This book was set in Adobe Jenson Pro and Frutiger LT Std by the author. Printed and bound in the United States of America.

Library of Congress Cataloging-in-Publication Data
Names: Smith, Brian Cantwell, author.
Title: The promise of artificial intelligence : reckoning and judgment / Brian Cantwell Smith.
Description: Cambridge, MA : The MIT Press, [2019] | Includes bibliographical references and index.
Identifiers: LCCN 2018060952 | ISBN 9780262043045 (hardcover : alk. paper)
Subjects: LCSH: Artificial intelligence--Philosophy. | Artificial intelligence--Social aspects.
Classification: LCC Q334.7 .S65 2019 | DDC 006.3--dc23 LC record available at https://lccn.loc.gov/2018060952

10   9   8   7   6   5   4   3   2

In memory of John Haugeland,
friend and fellow traveler

Contents

Preface

At lunch on March 25, 1980, at a workshop on Artificial Intelligence and Philosophy at Stanford's Center for Advanced Study in the Behavioral Sciences, one of the participants suggested that we would all have been better off if Kant had written his *Critique of Pure Reason* in Lisp.[1] Looking up in alarm, I locked eyes with an equally shocked participant, in the person of John Haugeland. So began a fast friendship and intellectual partnership that lasted until May 22, 2010, when John tragically succumbed to a heart attack at his own Festschrift at the University of Chicago.[2]

Haugeland was a philosopher, trained in existentialism and the philosophy of Heidegger, who devoted considerable resources to exploring the philosophical foundations of artificial intelligence (AI). I met him just as I was completing a doctorate in computer science at the Artificial Intelligence Laboratory at MIT. From that technical background I have spent much of my career reaching in the opposite direction from John, but toward the same territory: the philosophical imbrications of computation and mind. Our backgrounds were contrapuntal, but the common interest arose out of a deeper similarity: Haugeland's

---

1. I wrote my dissertation on Lisp; it was not for lack of familiarity with the programming language that I was appalled.
2. Haugeland lived for another month, passing away on June 23, but never regained consciousness after the conference.

father was a machinist;[3] my own, a theologian. Neither of us had in fact ever been purely one thing or the other. We grew up and dwelt in the same textured landscape between the technical and the philosophical—a landscape we were destined to spend hundreds of hours exploring together. "That's what I was just about to think!" became a staple of our endless conversations.

In the circumstance, I was unable to give a paper at Haugeland's Festschrift. This work is an attempt to lay out some of the views I would like to have presented.

Thanks especially to Güven Güzeldere, Amogha Sahu, and Henry Thompson for reading multiple drafts, and for extended ensuing discussions of these topics. Thanks as well to Christoph Becker, Jim Brink, Tony Chemero, Ron Chrisley, Nils Dahlbäck, Sandra Danilovic, Gillian Einstein, Lana el Sanyoura, Robert Gibbs, Vinod Goel, Stefan Heck, Steven Hockema, Atoosa Kasirzadeh, Jun Luo, Jim Mahoney, Tobias Rees, Murray Shanahan, Arnold Smith, and Rasmus Winther—as well as two anonymous reviewers. I also want to thank Ron Chrisley and the members of the "Minds, Machines, and Metaphysics" community (3M), first convened at the University of Indiana in the 1990s, that Ron has recently reconstituted, online, bringing together many of students with whom it has been my honor to work over the years. I know for a fact that no one mentioned here agrees with everything in this book, but I am nevertheless extremely grateful to them all. I can assure

---

3. While a graduate student, Haugeland built himself a printer by assembling a grid of relays, to which he attached bent paper clips that reached up and pulled down the keys on a scavenged IBM Selectric typewriter. He fabricated appropriate electronics and wrote a program to control the device, which served him for many years.

each and every one that the book is immensely better for their interventions.

Thanks to Robert Prior, Executive Editor at the MIT Press, for stewardship of this book and for faith and support over many years, and to Elizabeth Agresta for assistance in publication details.

I am also extremely grateful to Reid Hoffman for supporting the writing of this work, the research on which it is based, and the chance to pursue these topics over coming years. I also want to acknowledge his challenge to me, on December 16, 2017, to think about how we humans should approach living with synthetic forms of intelligence. I view this book as a chance to give voice to some initial thoughts.

Thanks beyond words to Gillian Einstein, as well—without whom I dare not imagine.

Introduction

Neither deep learning, nor other forms of second-wave AI, nor any proposals yet advanced for third-wave, will lead to genuine intelligence. Systems currently being imagined will achieve formidable reckoning prowess, but human-level intelligence and judgment, honed over millennia, is of a different order. It requires "getting up out" of internal representations and being committed to the world *as world*, in all its unutterable richness. Only with existential commitment, genuine stakes, and passionate resolve to hold things accountable to being in the world can a system (human or machine) genuinely refer to an object, assess ontological schemes, distinguish truth from falsity, respond appropriately to context, and shoulder responsibility.

Does that doom AI? No. Automated reckoning systems will transform human existence. But to understand their capacities, liabilities, impacts, and ethics, and to understand what assemblages of people and machines should be assigned what kinds of task, we need to understand what intelligence is, what AI has accomplished, and what kinds of work require what kinds of capacity. Only with a carefully delineated map can we wisely choreograph the world we are developing—the world we will jointly inhabit.

This book is intended as a contribution to that cartographic project. It is written out of a belief that the rise of computing and AI is of epochal significance, likely to be as consequential as the Scientific Revolution—an upheaval that will profoundly alter our understanding of the world, ourselves, and our (and our AIs') place in that world. So encompassing is the reconfiguration being catalyzed by computing and AI that we need fundamentally new ontological and epistemic frameworks to come to terms with it—new resources with which to ask such ultimate questions as *who we are, who to be, what to stand for, how to live.*

With an eye toward such questions, this book develops some intellectual tools with which to assess current developments in AI—especially recent advances in deep learning and second-wave AI that have led to such excitement, anxiety, and debate. The book will not assess specific projects, or recommend incremental advances. Instead, I will adopt a general strategy, in pursuit of two aims:

1. To unpack the notion of intelligence itself, in order to understand what sort(s) we humans have, what sort(s) AI aims at, what AI has accomplished so far, what can be expected in the foreseeable future, and what sorts of tasks can responsibly be assigned to systems currently constructed or imagined.
2. To develop a better understanding of the underlying nature of the world, to which I believe all forms of intelligence are ultimately accountable.

The second concern gives the book a strongly ontological flavor. In the end, I argue (i) that the deepest reason for the failure of first-wave AI was the untenable ontological world view on which it was founded, (ii) that the most important insight of second-wave AI is the window it gives us onto an alternative ontological perspective, and (iii) that the nature of reality implies that constructing anything warranting the label "artificial general intelligence" (AGI) will require developments far beyond and quite different from those imagined in either first- or second-wave AI,[1] involving various forms of committed, participatory engagement with the world.

I use **judgment** for the normative ideal to which I argue we should hold full-blooded human intelligence—a form of dispassionate[2] deliberative thought, grounded in ethical commitment and responsible action, appropriate to the situation in which it is deployed. Not every human cognitive act meets this ideal—not even every conscious act of every person we call intelligent. I claim only that judgment is the standard to which human thinking must ultimately aspire.[3]

Judgment of this sort, I believe, is a capacity we strive to instill in our children, and a principle to which we hold adults accountable. It is an achievement that far transcends individuals—a resource that has been forged, over

---

1. Or any that have been proposed for third-wave AI.

2. "Dispassionate" (and "disinterested" when I use the term) in its original sense of being fair, unbiased, open-minded, and free of prejudice. I neither mean nor intend to suggest that judgment should (or can) lack in care or commitment. On the contrary, as argued in chapter 11 and again in chapter 13, judgment must be simultaneously passionate, dispassionate, and compassionate.

3. This is not to suggest that judgment, as a regulative ideal, is remote from either consciousness or experience. One of the most important accomplishments of the historical development of culture, I believe, in diverse forms around the world, is to have established standards of judgment as a background condition on what it is to be a responsible adult. The fact that contemporary rending of the fabric of public discourse (perhaps abetted by digital technologies) is so widely decried stands witness to the fact that such norms have not been forgotten, even if they appear to be under threat.

Coming to understand what it is or would be to hold the human condition accountable to such an ideal is an added benefit of documenting what AI systems are, and what they are not.

thousands of years and in diverse cultures,[4] as a foundation for rationality, thought, and deliberative action, into which individuals must be recruited. It need not be articulate, "rationalistic," or independent of creativity, compassion, and generosity—failings of which (especially formal) logic is often accused. Rather, by judgment I mean something like what we get at (or should be getting at) when we say that someone "has good judgment": a form of thinking that is reliable, just,[5] and committed—to truth, to the world as it is.

With judgment in view as the ultimate goal of general intelligence, I examine the history of artificial intelligence, from its first-wave origins in what Haugeland dubbed "Good Old Fashioned AI" (GOFAI) to such contemporary second-wave approaches as deep learning. My aim is neither to promote nor to criticize—but to understand. What assumptions underlie the various technologies we have constructed? What conceptions of intelligence have been targeted at each stage? What kinds of success have been achieved so far, and what can be expected in the future? What aspects of judgment will contemporary AI systems reach, and what aspects have they not yet begun to

---

4. The development of judgment, we can safely presume, was accomplished without requiring alteration in DNA or neural architecture.
5. Philosophical readers may balk at the inclusion of justice and ethics in a norm on truth. As will become increasingly clear, I take *accountability to the world* to be something of an "ur-norm" that underlies not only truth but ethics, care, and compassion as well. To see how and why that is true, however, requires understanding what the "world" is, how ontology and truth arise, how existential commitment is a precondition for ontology, and so on. An argument for this metaphysical position is beyond the compass of this book; I will take up that project elsewhere, but for an initial glimpse see my *On the Origin of Objects* (Cambridge, MA: MIT Press, 1996).

approach? And to up the ante, and in order to bring into view one of the most important issues facing us today: can articulating a conception of judgment provide us with any inspiration on how we might use the advent of AI to raise the standards on what it is to be human?

I use the term **reckoning** for the types of calculative prowess at which computer and AI systems already excel—skills of extraordinary utility and importance, on which there is every reason to suppose computers will continue to advance (ultimately far surpassing us in many cases, where they do not do so already), but skills embodied in devices that lack the ethical commitment, deep contextual awareness, and ontological sensitivity of judgment. The difference between reckoning and judgment, which I will argue to be profound, highlights the need for a textured map of intelligence's kinds[6]—a map in terms of which to explain why reckoning systems are so astonishingly powerful in some respects, yet fall so spectacularly short in others.

Four caveats. First, the book is not a comparison of humans and machines. I see no reason to doubt that it may someday be possible to construct synthetic computational systems capable of genuine judgment. Or perhaps equivalently: if, as may happen, we construct synthetic creatures capable of evolving their own civilizations, or of incrementally participating in ours, nothing in my argument

---

6. Psychology has developed detailed conceptual maps of the human psyche, distinguishing such capacities as cognition, sensation, memory, and so on. As noted in chapter 2, the conception of intelligence on which AI was founded was very general, not making any such theoretical distinctions. I believe the kind of map we need in order to assess AI, moreover, is of a different order from that provided by psychology—in part because computers are opening up vast regions unoccupied by humans or nonhuman animals.

militates against the possibility that, in due course, such creatures might themselves evolve judgment—much as we have, and perhaps with no more explicit understanding of their capacities than we have of our own. Nor am I arguing that cyborgs or other human-machine assemblages, *for that reason alone*, will be challenged in regard to their capacity for authentic judgment. My claims are just two: (i) the systems we are currently designing and building are nowhere near that point; and (ii) no historical or current approaches to AI, nor any I see on the horizon, have even begun to wrestle with the question of what constructing or developing judgment would involve.

Yet attempting to reach this conclusion by drawing a distinction between DNA- and silicon-based creatures would be a grave mistake, in my view—chauvinist, sentimental, and fatally shallow. Rigor demands that we articulate a space of possible kinds of intelligence in terms of which AIs, humans, and nonhuman animals can all be evenly and nonprejudicially assessed.

Second, as will progressively emerge, judgment in the sense I am defending it is an overarching, systemic capacity or commitment, involving the whole commitment of a whole system to a whole world. I do not see it as an isolable property of individual capacities; nor do I believe it is likely to arise from any particular architectural feature—including any architectural feature missing in current designs. Readers should not expect to find specific architectural suggestions here, or recommendations for technical repair. The issues are deeper, and the stakes higher, than can be reached by any such approach.

Third, I am fully aware that the conception of judgment I will defend does not fit into any standard division between "rational thought," on the one hand, and "emotion" or "affect," on the other. On the contrary, one of my aims is to unsettle reigning understandings of rationality, in part to break it up into different kinds, but also to suggest that reason in its fullest sense—reason of any sort to which we should aspire—necessarily incorporates some of the commitments and compulsions to action that are often associated with affectual states. These moves arise out of a larger commitment: if we are to give the prospect of AI the importance it deserves, we must not assume that time-honored conceptions of rationality will survive unscathed.

Fourth, although I take seriously many of the critiques of first-wave AI articulated in the 1970s, this book is by no means intended to be an updated treatise along the lines of Dreyfus's *What Computers Can't Do*.[7] On the contrary, one of my goals is to develop conceptual resources in terms of which to understand what computers *can* do. In fact, the entire discussion is intended to be positive. I do not plead for halting the development of AI, or argue that we should bar AI systems from functioning in situations of moral gravity. (When landing in San Francisco, I am glad the airplane is guided by sophisticated computer systems, rather than by pilots peering out through the fog looking for the airport.) I am also not worried, at least here, about whether AI systems will grow more powerful than we humans, or that they will develop their own consciousness. And I take seriously the fact that we will soon need to learn how to live in productive communion with synthetic intelligent creatures of our own (and ultimately their) design.

Two things do terrify me, though: (i) that we will rely on reckoning systems in situations that require genuine judgment; and (ii) that, by being unduly impressed by

---

7. Hubert Dreyfus, *What Computers Can't Do: A Critique of Artificial Reason* (New York: Harper & Row, 1972).

reckoning prowess, we will shift our expectations on human mental activity in a reckoning direction. Current events give me pause in both respects. The calls to which I believe we should respond, and to which I hope this book will draw our attention, are: (i) that we learn how to use AI systems to shoulder the reckoning tasks at which they excel, and not for other tasks beyond their capacity; and (ii) that we strengthen, rather than weaken, our commitment to judgment, dispassion, ethics, and the world.

1 — Background

There was excitement the first time too. In the 1960s and 1970s artificial intelligence was in its first flush. Computers' stunning power had only recently come into focus, electrifying the newly minted AI labs. What gripped our imagination was not just computing's potential impact on society, impressive as that was, but the thought that we humans, too, might be computers. The idea unleashed visions of grandeur. If we could just "program things up," we dreamed, we could put paid to thousands of years of philosophy, surround ourselves with intelligent companions, and understand the human condition.

It is easy to disparage that bravado now. Those "first-wave AI" systems were brittle, unconducive to learning, defeated by uncertainty, and unable to cope with the world's rough and tumble. Yet there were also successes, many of which paved the way for what is now commonplace—online commerce, navigation systems, social media. But in a long-standing pattern, once tasks were automated they no longer seemed to require so much intelligence. Gradually the ebullience waned.

There is excitement once again—for "second-wave" AI, this time, for a new "AI spring." Deep learning and affiliated statistical methods, backed by unimagined computational power and reams of Big Data, are achieving levels of performance that trounce the achievements of the earlier phase. The profession is psyched, the commercial world scrambling, the press giddy with anticipation. But there is also anxiety. What captures our imagination is not so much awe, this time, as *consequence*. Exhilaration is being replaced by terror—that new and perhaps alien AIs are going to take over our jobs, our lives, our world.

Are we genuinely on the cusp of what we only prematurely imagined in the 1970s? Or is this, too, a phase destined to pass? One thing is sure: there is no dearth of opinions. Testimonials about second-wave AI flood the web, along with comparisons with first-wave, and, as new limitations come into view, proposals for "third-wave" AI—for what it will take to achieve the ultimate goal of "artificial general intelligence" (AGI). There has certainly been astounding progress. World-beating Go programs, serviceable machine translation, and uncanny image recognition are genuinely stunning. Respected scientists are serious about the magnitude of the impending upheaval—the prophets of doom as well as the inveterate triumphalists.

But respected scientists were serious 50 years ago, too.

To take stock of what is happening, we need to understand the ontological, epistemological, and existential assumptions on which each wave of AI has been based. Viewing things historically will help us judge the capacities, implications, and limitations of the first two stages, and assess prospects for a third. It will also allow for a more discerning analysis than the standard characterizations of first-wave AI as based on "handcrafted symbolic representations" and second-wave on "statistical pattern matching over large data sets." Taking the measure of each, and soberly assaying the future, requires something deeper: not just a unifying architectural perspective that subsumes both options (and perhaps others as well), but, more broadly, an understanding of what intelligence is, on the one hand, and what the world is like, on the other. It is time to channel McCulloch:[1] *what is the world, such that we can understand it—and who are we, such that we can understand the world?*

Three methodological preliminaries. First, as already indicated, it is imperative not to frame the debate in terms of *human* and *machine*. The problem is not just that those labels are vast, vague, and emotionally charged. More seriously, to avoid ideology and prejudice, we need independent evaluative criteria. To judge what kind(s) of intelligence current AIs have, what kind(s) we humans have, and what kinds are on offer for both machines and people, we need an understanding of "intelligence" and its "kinds" not circularly defined in terms of the entities to which we want to apply it.

Second, it is important to deflect potential confusion due to terminological differences between computer science and adjacent fields. A variety of classic terms from logic, philosophy, linguistics, and epistemology, having to do with relation of an intentional (representational, meaningful) system to the wider world, have been repurposed in computational parlance to refer to causal behaviors and arrangements within the confines of the machine itself. Especially important are *meaning, semantics, reference*, and *interpretation*. Consider the phrase "the semantics of program P." Many (including I) would take this term to refer to a relation between P and the world or task domain that P is "about" (that its data structures represent, in which it is deployed, about which it performs calculations). But from a perspective I call **blanket mechanism**, which has largely

---

1. Warren McCulloch, "What is a Number, that a Man May Know It, and a Man, that He May Know a Number?," *General Semantics Bulletin*, no. 26/27 (1960): 7–18.

engulfed computer science, the phrase is used to refer to the behavioral consequences, within the computer system, of P's being executed. In computer science, that is, the term "semantics" no longer takes one outside the boundaries of a mechanistic device. Since my concerns in this book are with the very "into the real world" relations that the terms traditionally referred to, I will use them here in their classic sense, marking those points where confusion is most likely to arise.[2]

Third, the gravity of the stakes suggests addressing the issues at a relatively high level of generality. Some questions are foundational: What is intelligence? What are its physical limits? What lies beyond the line of possibility, so that we can give up longing for it? And on this side: Where are we, now, and where are our creations? How will we want—how should we want—to live with other forms of intelligence, including not just products of natural evolution,

---

2. Not a few computer scientists have suggested, over the years, that I "get with the program" and realize that these internal mechanical relations and behaviors are what the terms *semantics, reference, interpretation*, and the like now mean. But the suggestion raises numerous difficulties. First, I *need* terms that mean what these terms classically meant. If I were to accept the redefinitions, or (more plausibly) were to avoid them in order not to sow confusion, I would lose any way to talk about what I do want to address. Second, I believe that retaining the classical meanings is necessary in order to understand the relations among AI, computing, mind, language, and thought. Third, even scientifically, I believe we need to retain the classical meanings in order to get at technical issues within computing itself (see my *Computational Reflections*, forthcoming). Fourth, and most importantly, I have no interest in supporting blanket mechanism in general, which I see infecting wider and wider swaths of intellectual inquiry. On the contrary, I believe it is vital that we not submit to the rising tide of entirely mechanistic world views and explanations. So the redefinitions and repurposings are moves I want to resist.

but entities of our own devising, and eventually of theirs as well? What impact will and should the development of synthetic[3] intelligence have on human intelligence, on our sense of self, on standards for humanity?

There are also practical questions. What sorts of intelligence are required to perform different sorts of task—driving a car in the city, reading an X-ray, building houses, teaching children, detecting racism, routing propaganda and fake news? How can we responsibly assign specific kinds of work to appropriate configurations of people, machines, and processes? What divisions of labor, between and among organisms and devices and communities and systems and governments, are sustainable, honorable—and humane?

These questions target the future. To address them, we need a grip on the past.

---

3. A better word than "artificial," if it can ever be achieved.

2 — History

It is not random that classical first-wave AI was based on symbolic representation. Its approach, immortalized in Haugeland's indelible phrase "good old-fashioned Artificial Intelligence" (GOFAI),[1] grew out of four vaguely Cartesian assumptions:

C1 The essence of intelligence is **thought**, meaning roughly rational deliberation.

C2 The ideal model of thought is **logical inference** (based on "clear and distinct" concepts, of the sort we associate with discrete words).

C3 **Perception** is at a lower level than thought, and will not be that conceptually demanding.[2]

C4 The **ontology** of the world is what I will call *formal*:[3] discrete, well-defined, mesoscale objects

---

1. John Haugeland, *Artificial Intelligence: The Very Idea* (Cambridge, MA: MIT Press, 1985), 112.

2. It is not that anyone believed that perception would be *simple*. Rather, the conceit was that perception, perhaps because so well exemplified in nonhuman animals, should not only be viewed as lower-level than real intelligence, but also, and crucially, much less mysterious—not of the same order as the really profound question of what genuine intelligence was, how it had emerged, how thought was possible in a mere physical device, etc. (That we are perceiving creatures was never thought to be an argument for dualism, or a reason to doubt our material constitution.)

3. This is by no means a standard definition of formality. The term is used in a variety of ways, to mean syntactic, abstract, and

exemplifying properties and standing in unambiguous relations.

Regarding C1, when AI was founded, the term "intelligence" was given no very specific meaning. The aim was simply to construct a machine that intuitively seemed "smart"—that could think, emulate human rationality, or demonstrate cognitive prowess. Since that time, people have substantially enlarged the scope of mental activity toward which they direct AI—most obviously by including perception, action, and categorization within mainstream AI research, but also by exploring emotion, cognitive development, the nature of assertion and denial, and so forth. But in the beginning, conceptual representation was largely taken as given, and the focus was directed toward a roughly logicist conception of thinking and reasoning.

There is nothing specifically computational about assumptions C1–C4.[4] An additional insight was required in order to tie them into computation, arising from nineteenth- and twentieth-century developments in logic in the work of Boole, Peirce, Frege, and others. It bears a word of explication here, partly because appreciation of it has waned,[5] but also because it allows us to frame the challenge that all intelligence must face.

···——————————

mathematical, among other things. See page 324 of *On the Origin of Objects* for an analysis, along the lines suggested here, of the common thread underlying all of these readings—of formality, *au fond*, as "discreteness run amok."

4. If anything, the association of computing with this model of intelligence ran the other way: Turing's proposal for his eponymous machines was based on his logicist conception of intelligence.

5. Though contemporary programmers may not recognize the formulation spelled out on the following pages, Turing would have found it unexceptional.

The insight rests on a fundamental thesis about representation and the nature of physical causality. It takes the form of a claim that it is possible to build a system with four crucial properties, which together allow (us to take) it to "reason" or "process information":

P1   The system **works**, mechanically, in ways explicable by science. We can build such devices. Nothing spooky is required. No magic or divine "inspiration"; no elixir of life or soul.

P2   The system's behavior and constituents[6] support **semantic interpretation**—that is, can be taken to be about (mean, represent, etc.) facts and situations in the outside world.[7]

---

6. Philosophers draw a distinction between a system's behavior or identity as an integrated whole, and its internal constituents (data structures, ingredients, etc.)—with properties of the former, in the human case, being called *personal*; the latter, *subpersonal*. The difference is easiest to understand in ethical cases. While we might hold you accountable for doing something egregious, if neuroscientists were to trace the reason for your misbehavior to abnormal functioning of your brain, it would be odd to say "Bad amygdala!" But the distinction between personal and subpersonal is hard to draw carefully, even in the human case. Are memories, which clearly admit of semantic interpretation—in fact are comprehensible only in such terms— constituents of your mind, or external (personal-level) behaviors?

The distinction matters greatly in some cases—including with respect to the notion of judgment being developed here, which I will argue to be an irrevocably "personal" or "systemic" capacity, rather than being the capacity of any ingredient or subpersonal part. It is not, however, relevant to this origin story about AI.

7. As mentioned in chapter 1, this is not what "semantics" means in computer science, where the term has come to be used as a name for the mechanically individuated behavioral consequences of a program or data structure. For details see my forthcoming *Computational Reflections*. More generally, the claim that computation is a semantic or

From these two properties it follows that a wedge opens up between two ways of understanding computers: (i) *what they do*, when that is understood "under interpretation"—that is, in terms of the entities in the world that the representations signify; and (ii) *how they work*, as (uninterpreted) causal mechanisms. Examples of the former (what they do under interpretation) include calculating the prime factors of a large number, figuring out the shortest route through all American state capitals, predicting the outcome of the fall election, telling time. These activities are not described in terms of what is going on inside the machine, but in terms of *what the world is or might be like*, based on an interpretation[8] of the machine's output or behavior. Predicting that the government will fall is a prediction about politics, not (in the first instance) about the characteristics of an output or data structure. Information that Montreal is north of Ottawa has to do with how the world is arranged, not with bits or channels or measures or symbols.

Understanding intelligence under interpretation is utterly commonplace.[9] If I ask "what's on your mind?" you are likely to reply "I'm thinking that _____," where the blank is filled with a description of the world, not of your thoughts as uninterpreted causal patternings (few are the people who would say "I have just increased

---

intentional phenomenon at all is increasingly challenged (see, e.g., Gualtiero Piccinini, *Physical Computation: A Mechanistic Account*, Oxford: Oxford University Press, 2015), though I stand by my claim.

8. "Interpretation" is another term that is given an internal, mechanist reading in computer science, but that I use here in its normal English sense, as having to do with the world or intentional subject matter.

9. In fact intelligence only *is* intelligence under interpretation. "Being intelligent" is not a description of uninterpreted causal behavior or mechanical configuration.

my cortisol level by 2.7% and sent 10^7 neural firings across my corpus callosum").[10]

Principles P1 and P2 are recognized as fundamental to the structure of logic, and underlie the "formal symbol manipulation" construal of computing as well. While I believe the next two also apply to both logic and computing, they are less commonly highlighted or remarked.

P3 The system is **normatively** assessed or governed[11] in terms of the semantic interpretation.

What matters about such systems—whether they are right or wrong, true or false, useful or useless—has to do with an assessment of what they are interpreted as representing or saying. Is it true that if we launch in that direction we will ultimately get to Saturn? That there are 78,498 primes less than a million? That at this rate it will take 87 years to pay off our mortgage? It does not matter how efficient an algorithm is if it produces a misleading or wrong answer.

---

10. I will have more to say about semantic interpretability later. For now, it is important merely to note that the issues of whether the interpretations are arbitrarily attributed and thus changeable at whim (as Searle presumes in his Chinese Room example), whether they are anchored or grounded in the world (in what is known as the "symbol grounding problem"), and whether the semantics is "original" or "derivative" (or "authentic" and "inauthentic," as discussed by Searle and Haugeland) are at least partially independent. See chapter 7, pages 78–79.

11. I would say "normatively evaluated," except that, like "semantics," the term "evaluate" has come to take on a purely behavioral meaning in computer science, at odds with its historically normative sense of worth, propriety, being "right" or true, etc.

By "governed" I do not mean causally or mechanically controlled, but something closer to what is meant by saying that a democratic society is "ultimately governed by respect and trust"—or that physical events are "governed by the laws of nature."

The fourth point is the most difficult to understand, but the most metaphysically basic. It raises a fundamental challenge for all forms of intelligence. I would even go so far as to say that it is what *establishes the problem that intelligence is "designed" to solve*. I will frame it in terms of **effectiveness**, meaning "something that can be done by physical or mechanical means." Setting aside complexities due to mathematical modeling, this is the sense that computer science studies under its label of "*effective* computability." Subtleties abound, but for present purposes the word can be understood as an approximate synonym for "immediately causal."[12]

P4 In general, semantic relations[13] to the world
      (including reference) are **not effective**.

The presence of a semantic relation cannot be causally detected at either end—neither at the signifying object or event (term, word, representation, data structure, thought, etc.) nor at the signified entity (reference, denotation, or represented entity or state of affairs). You cannot tell what a representation represents, for example, by local measurement or detection. No physical detector, located at an

---

12. Technically, "effective" should be understood as a higher-order property, holding of those properties in virtue of the exemplification of which something can do physical (causal) work. That said, the relation between computer science's notion of effectiveness and classical treatments of causality warrants investigation—especially once it is recognized that the notion of effective computability cannot be defined purely abstractly, which I believe. See my "Solving the Halting Problem, and Other Skullduggery in the Foundations of Computing," unpublished manuscript.

13. By *semantic relations* I mean the relations between signs (terms, noun phrases, representations, descriptions, thoughts, data structures, etc.) and the objects that they signify (refer to, denote, describe, represent, or are about).

object, can determine whether that object is the object of a description or directed thought. While "being referred to or represented" is a perfectly real property (which often matters a great deal), no wave of discriminable energy travels along the arrow of intentional directedness; it is not a signal that any physical instrument can pick up. Creatures on Andromeda could be thinking about us right now, without there being any way for us to know or find out; similarly, you can refer to the sun without eight seconds having to pass before the sun is referred to. Nor can reference (aboutness) be blocked by physical barriers. I can think about Alpha Centauri even if you lock me in a lead vault. If, in that vault, I suppose that Reno is east of Los Angeles, I can still be wrong.[14]

The noneffectiveness of reference and semantics[15] is fundamental to the nature of representation and intelligence,

---

14. I describe the following iPhone application to my students, guaranteeing—with my life!—that if they can build it, they will earn a billion dollars. The application is specified to work as follows: once someone loads it onto their phone and starts it up, the phone will beep every time someone thinks about them.

Many students, being committed mechanists, claim that they could surely build such an app, at least conceptually, by instrumenting everyone's brains, uploading everyone's thoughts to the cloud, sending a signal to the target's phone at just the right time, and so forth. But that is not the point—nor the challenge. My claim is that there is a fact of the matter, *right now*, as to whether any particular person is being thought about at any given instant—but that fact does not result in any physical disturbance or "signal" arriving at that person.

I am not worried.

15. In saying that semantics is not effective, I am not denying that semantical acts can and do have causal consequences. "Can you reach the salt?" may trigger the salt's being passed; Galileo's defense of heliocentrism led him to be branded as a heretic. The point is that intentional acts neither make nor require effective (causal) contact with the states of affairs they are about.

and radically affects ontology as well—the objects and properties in terms of which we find the world intelligible. It is a consequence of the most basic fact in all of physics: that causal influences fall away locally in both space and time. The fact that the universe is a web of purely local influence poses a formidable challenge to any system or creature aiming to comprehend its world. It means that no system, human or machine, can determine what is going on merely by "looking out" and seeing or sampling it (drastically limiting the applicability of Brooks's famous maxim that "the world is its own best model"[16]). Most of what matters, most of what an agent cares about—everything in the past, everything in the future, everything that is far away, and a huge number of facts about what is close by, too[17]—cannot be directly detected. Whether the news is true, whether the chair needs to be moved inside, whether you will ever come to appreciate the advice you were given—none of these things can simply be "read off" the impinging field of causal disturbance.

Yet even if we cannot directly detect it, we need to know—if we are intelligent—what is going on, distally: where we stored those acorns, that a predator lurks behind that rock, that the person who just left the room still exists, rather than having metaphysically vanished, that the sun will rise again tomorrow. If intelligence involves finding

16. Rodney Brooks, "Intelligence Without Reason," MIT Artificial Intelligence Laboratory Memo 1293 (1991); the phrase also occurs in the version of his "Intelligence Without Representation" in John Haugeland, ed., *Mind Design II: Philosophy, Psychology, Artificial Intelligence* (Cambridge, MA: MIT Press, A Bradford Book, 1997), p. 405. The point in the text is not that Brooks's thesis is false; it is that it is only useful with respect to immediately surrounding circumstances.
17. All those states of affairs that involve the exemplification of non-effective properties.

the world as a whole intelligible, which I assume to be the case, but we are in effective contact with only that minuscule fraction that is nearby (the effective facts within a $1/r^2$ spatiotemporal envelope), then it is fundamental to intelligence to understand what is going on beyond what is effectively proximal. Put it more strongly: the world's being beyond effective reach is *the only reason we need reasoning at all.* As Strawson pointed out, we need to "know that our senses fail, rather than that the world fades."[18]

How do we orient appropriately to a world beyond effective reach? By making use of locally effective structures and processes in the world's stead—that is, by **representing** the world. This is essentially a metaphysical requirement on intelligence.[19] That it does not contravene

---

18. P. F. Strawson, *Individuals* (London: Methuen, 1959). Similarly, "We think so that our hypotheses can die in our stead," Popper's famous maxim, makes sense only because of the discrepancy between the semantic reach of our thoughts and the causal extent of the world.

19. Representation is widely decried, these days, throughout cognitive science and much of the humanities. These antirepresentationalism sentiments are arrayed against a very narrow notion of representation, however, thereby discarding one of the deepest insights into what reasoning and intelligence are. Acknowledging the tension between what is effectively proximal and what is normatively pertinent but physically distal does not require treating representation as necessarily symbolic, pledging allegiance to naive realism, denying that ontology depends on culture or practice, asserting that cognitive activities are exclusively rational or deliberative, affirming a strict correspondence between the structure of the sign and the structure of the signified (a common interpretation of the Wittgensteinian "picture theory" of representation), or a host of other presumptive ills. Representation as I am characterizing it here is as pertinent to enactive, feminist, poststructuralist, and ecological views of cognition as to those traditionally labeled as intentional or representational. See my "Rehabilitating Representation" (unpublished manuscript presented at the Center for Cognitive Science, University at Buffalo, State University of New York, Spring Colloquium, 2008).

P4 (the noneffectiveness of semantics) stems from the fact that representations have two criterial properties: (i) they stand in noneffective semantic relations with what is distal, with what is not immediately effectively available (what they represent); and (ii) they can nevertheless effect or be affected by physical changes within causal proximity. Qua organized pattern of ink on a surface, a sign inscribed with "INTERSECTION AHEAD" can reflect light differentially, causing signals to be activated in the retina, in turn triggering neural activity causing braking behavior; at the same time, within the regime in which it functions as a representation, the sign can stand in some kind of structural relationship (fashioned in such a way as to normatively guide action) with a state of affairs out of sight around the corner. So too a thought about an upcoming speed trap can helpfully cause you to slow down before the police can detect your approaching vehicle—that is, before you are within effective reach of the radar.

The use of representations with these coordinated dual properties is what allows an intelligent system to be physically possible (P1). But in order for such symbols and thoughts about what is beyond effective reach to be *correct* or *worthwhile*, those "within reach" local, effective structures and the processes defined over them must be governed by **norms** framed in terms of the distal, "out of reach" (noneffectively available) situations that the system is about.[20] To put it in a single sentence:

---

20. Soundness and completeness are the specific versions of these norms appropriate to the particular assumptions and restrictions applicable to formal logic.

◆ REPRESENTATIONAL MANDATE: *The proper functioning of any world-directed system—any system that
is thinking about or representing or processing information about the world—must be governed*[21] *by normative criteria (P3) applying to its mechanical operations
(P1) that are framed in terms of situations and states
of affairs in the world that the system is representing
or reasoning about (P2), which situations and states of
affairs will not, in the general case, be within effective
(causal) reach (P4).*

Because the norms are framed in terms of the world, and
govern the local mechanism—that is, because what matters, in order for the reasoning to be proper or worthy, is
what is going on in the world—I say that reasoning systems, and all of intelligence, must be **deferential**.

The impact on ontology is direct. Representations can
satisfy distal norms (norms defined in terms of distal and
noneffectively accessible states of affairs) only if their local, effective properties correspond appropriately[22] to the
properties of the represented situations in whose stead
they serve. It follows that the world, to be intelligible, must
manifest some degree of time- and distance-spanning correlation.[23] At any given place and moment in time, smoke
can be recognized as a signal of fire only if the causally efficacious (i.e., effective) presence of smoke reliably correlates

---

21. See note 11 (p. 11).
22. "Appropriate correspondence" may include structures that medi-
ate activity—or even such activity itself—so long as it leads to behavior normatively appropriate to that which is thereby represented.
23. I take the quantitative, nonsemantic conceptions of information
espoused in computer science and physics to be (postontological)
measures of the world's correlation. I do not believe such correlations
should be called *information*, but they are undeniably important.

with the (potentially unobservable, and thus at least potentially not causally efficacious) fact of something burning,
perhaps distally. If reality was everywhere independent,
the universe would be random, representation impossible,
and intelligence nonexistent. At the same time, the world is
not completely correlated; that would lock it up into a rigid clutch, with representation, intelligence, and creatures
again proscribed.[24] At any level at which we find it intelligible, the world must be *partially connected* and *partially
disconnected*.[25]

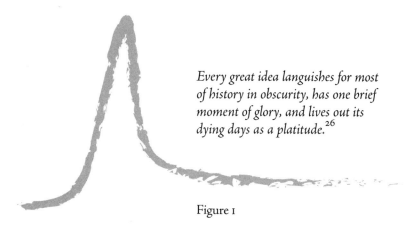

*Every great idea languishes for most
of history in obscurity, has one brief
moment of glory, and lives out its
dying days as a platitude.*[26]

Figure 1

24. See the discussion of the "gear world" in *On the Origin of Objects*,
199–200.
25. Some may be tempted to phrase this point in terms of degrees
of freedom, but any such measure requires ontology, and ontology
rests on this pattern of partial connection and disconnection, not the
other way around. It might merit investigation, though, to explore
what this "balance" between freedom and correlation that is so crucial
to registration, information, finding the world intelligible, etc., is due
to physical laws, constitutive conditions governing the abstractions
we employ in our registrations, or other factor(s).
26. I heard this saying from my father when I was a child. Whether it
was original to him or whether he was quoting someone else I do not
know, but I have been unable to determine another source.

The patterns of connection and disconnection necessary for representation, and hence for intelligence, in turn
put conditions on possible ontologies. It is no accident that
your personhood is not contingent on the moment-to-mo-
ment orientation of your limbs—that you do not go out
of existence when you stand up, with a new person magically springing into being. If that were true—if our concept
of "person" constrained limb orientation—I would not be
able to know, at any given moment, whether you existed.
In fact no one[27] would have a continuing identity at all. It
is not just that intelligence requires the ability to track and
think about objects, in other words; *objects must be patches
of reality that can be thought about and distally tracked.* Epistemology constrains ontology; it is not just the other way
around.[28]

That thinking, intelligence, and information processing
satisfy criteria P1–P4 is the fundamental idea behind formal logic and computing. That they are not just things
that humans do, but also things we can build automatic

---

27. That is, no space-time worm that we, in our current ontology (registration scheme; pp. 35–36), register as a person.

28. Realists need not panic. The point is not that epistemological or
psychological states somehow magically reach out and mysteriously
adjust the metaphysical plenum. Nor is it merely that registrational
practices are genuine activities in the world (though they are). Rather,
as I put it below, our ontological (or as I put it, "registration") schemes
must be such as to support epistemic access to their objects—must
parse reality in such a way that the resulting "objects" can be referred
to and thought about. What it is to be an object, in other words, is
to be a possible locus of registrational accountability. Yes, an object is
different from our representation of it; but that does not imply (or
require) that ontology and epistemology are logically independent.

   I am using "constrain," here, along the lines of the "govern" in the
REPRESENTATIONAL MANDATE (p. 17).

machines to do, was a stunning discovery in the first several decades of the twentieth century. By now it is so obvious that it barely garners mention (see figure 1, page 18). But the insight is fundamental—and, being a precondition for the very possibility of intelligence, impossible to imagine not being true. It enabled the study of formal logic, underwrote the development of the computer, unleashed the first generation of AI systems, led to the development of knowledge representation languages, and in general shaped our conception of how to build world-directed computer systems.

In a moment I will argue that this general insight underlies second-wave AI, too. The initial idea about how to manifest it in a machine, however—that is, the conception underpinning GOFAI or first-wave AI (I will use the two terms essentially interchangeably)—was more limited than initially realized. In the GOFAI era, it was assumed that the evident way to implement it was to build an interconnected network of discrete symbols or data structures, organized in roughly propositional form, operating according to a regimen dictated by an equally symbolic program. This classical "symbolic" architecture still underlies such staple programs as health record systems, Facebook databases, and the calculations of the Internal Revenue Service. There is a sense in which it is still in use in all programs (such as Microsoft Word).

◆   ◆   ◆

To a contemporary imagination, especially one entranced by the achievements of second-wave AI, the intellectual underpinnings of GOFAI and first-wave AI may not seem all that impressive—certainly not as groundbreaking as they did in the 1960s and 1970s. But it is essential to recognize that not only did they launch the whole AI enterprise, and in fact the entire development of computing, but the insights on which they are based remain durably deep and powerful. If we are to draw a comprehensive map of the territory of AI, they need to be given appropriate and respectful place.

3 — Failure

Nevertheless, GOFAI failed. Or anyway it is deemed to have failed.[1] The failures can be summarized as of four main types.[2]

F1 **Neurological:** The brain does not work the way that GOFAI does. All of today's mainstream CPUs, and all classical AI systems, consist of relatively few serial loci of activity (one to a few dozen threads) running at extreme speeds (on the order of 10^9 operations per second), pursuing relatively deep levels of inference or procedural nesting. So far as we know, the functioning of the brain is almost the complete opposite. Brain operations are roughly 50 million times slower, deriving their power from massive parallelism.[3]

---

1. GOFAI was not abandoned. Research in its vein has continued; in chapter 7 I argue that it addresses issues crucial to forward progress. But GOFAI no longer has imaginative grip as "how intelligence works"; it is widely disparaged in cognitive science and neuroscience; the center of gravity of AI thinking has shifted to machine learning and other second-wave approaches; and, as I argue here, a number of its founding assumptions (C1-C4) are untenable as a general model of intelligence.

2. This typology is loosely based on Part II of Dreyfus's landmark *What Computers Can't Do*, 1972.

3. According to what is known as Feldman's "100-step rule" (Jerome Feldman and Dana Ballard, "Connectionist Models and their Properties," *Cognitive Science* 6, no. 3, 1982, 206), the brain can compute at

Whether these low-level architectural differences are germane to the achievement of human-level intelligence, however, is not a straightforward question. To answer would require knowing what it takes to have intelligence like ours, what intelligence of a nonhuman kind might be like, what sorts of implementation strategies the brain uses, and so forth. Nevertheless, GOFAI's decline has led to renewed interest in "brain-style" computing, of which today's "neural networks," deep learning systems, and other machine learning architectures are instances.[4]

F2 **Perceptual:** Most GOFAI theorists thought that perception—recognizing or "parsing" the world based on perceptual input from sensors—would be conceptually simpler than simulating or creating "real intelligence." After all, in Descartes's phrase, "mere beasts" are good at it, in some cases notably better than people. Plus, the ontological assumption underlying GOFAI (C4; also F4, below), according to which the world consists of unambiguous objects exemplifying discrete properties, suggested that perception would merely involve determining what objects were "out there," what properties they manifested, what types or categories they belong to, and so on.

It did not work out that way. The difficulty of

··········

most 100 serial steps per second—sobering, given the brain's ability to perform many cognitive tasks in under a second. .

4. Neurally inspired architectures have a long history, going back to McCulloch and 0. And not all massively parallel architectures were neurally inspired—for example, some early semantic nets. But it is unarguable that the imaginative center of attention in AI has shifted from serial architectures to massively parallel networks.

Otto Lowe, 2018

Figure 2

real-world perception was one of several profound humilities to which we have been led by our experience with first-wave AI.

When people first attached digital cameras to computers, they were flabbergasted. This was the early 1970s, after all. The idea of using electronics instead of film had just been bruited (in 1961 in regard to satellites; the first digital camera was not released to the market until 1975, toward the end of GOFAI's reign). Why were people taken aback? Because what came through the sensors was a mess. It turns out that the idea that the world consists of a simple arrangement of straightforward objects is the delivery, to our consciousness, of the results of an exquisitely sensitive, finely tuned perceptual apparatus running on a 100-billion neuron device with 100 trillion interconnections honed over 500 million years of evolution. Figure 2 may seem like a straightforward picture of an artist[5] in a workshop; figure 3 is a rendering this artist once made to convey to human (visual) observers what he believes the world is actually like—that is, it is an image that, *after* human perceptual processing, gives us a rough sense of what the world looks like *prior* to human processing.

Adam Lowe, CR29, "Studio with phone," 1993[6]

Figure 3

5. Adam Lowe, of Factum Arte (http://www.factum-arte.com).
6. Collection of Adrian Cussins. I used this image as chapter 10 of *On the Origin of Objects*.

Descartes was no fool. His standards on rationality were much higher than commonly appreciated. But history suggests he was insufficiently impressed with the achievement of looking out and
seeing a tree.

F3 **Epistemological:** Thinking and intelligence, on
the GOFAI model, consisted of rational, articulated steps, on its founding model of logical inference.
As Dreyfus insisted, and many cognitive theorists
have emphasized,[7] in many situations a more accurate characterization of intelligence is one of skill-
ful coping or navigation—of being "thrown" into
individual and social projects in which we are embedded and enmeshed. Even thinking, these writ-
ers argue, rather than consisting in a series of read-
ily articulable steps, emerges from an unconscious
background—a horizon of ineffable knowledge
and sense-making.

---

7. For example, Joseph Weizenbaum, *Computer Power and Human
Reason: From Judgment to Calculation* (New York: W. H. Freeman
and Company, 1976); Lucy Suchman, *Human-Machine Reconfigu-
rations: Plans and Situated Actions* (Cambridge: Cambridge Uni-
versity Press, 2007); Humberto Maturana and Francisco Varela,
Autopoiesis and Cognition: The Realization of the Living (Dordrecht:
Reidel, 1980); Terry Winograd and Fernando Flores, *Understanding
Computers and Cognition: A New Foundation for Design* (Norwood,
MA: Ablex Publishing, 1986); Eleanor Rosch, Francisco Varela, and
Evan Thompson, *The Embodied Mind* (Cambridge, MA: MIT Press,
1991); Evan Thompson and Francisco Varela, "Radical Embodiment:
Neural Dynamics and Consciousness," *Trends in Cognitive Sciences* 5,
no. 10 (2001): 418–425; and Daniel Hutto, "Knowing What? Radical
Versus Conservative Enactivism," *Phenomenology and the Cognitive
Sciences* 4, no. 4 (2005): 389–405.

F4 **Ontological:** The misunderstanding about perception, and perhaps about thinking as well, betray a much deeper failure of first-wave AI: its assumption, already mentioned (C4), that the world comes chopped up into neat, ontologically discrete objects. In fact one of the primary theses of this book is that the misconception of the world in first-wave AI is the "smoking gun" that explains its ultimate inadequacy. In chapter 6, I will argue that the successes of second-wave AI can be interpreted as evidence for the inadequacy of formal ontology both as a basis for intelligence and as a model of what the world is actually like.[8]

Of these four types, the ontological issues cut the deepest. They not only underlie the perceptual and epistemological difficulties, but also need to be appreciated in order to understand how and why second-wave AI has made at least some progress toward interpreting the world, especially at the perceptual level.

Three points are especially relevant.

First is a general thesis with which few GOFAI designers would disagree, even if no GOFAI system could itself be said to understand it. Though framed epistemically (as having to do with intentionality), the point rests on underlying ontological facts. It is universally accepted that the world can be described or conceptualized in multiple ways—as is often said, at various "levels of description." Reality itself is assumed, at least for all practical purposes,

---

8. In chapter 7 I argue that preserving what was good about GOFAI requires extracting its insights from the presumption that they can be understood in terms of, and uniquely apply to, a world adequately characterized in terms of formal ontology.

to be surpassingly rich, about which any ontological "parse" provides only partial information. Representations, descriptions, models, and the like all interpret or picture or filter the world through abstractions or idealizations— conceptual "frames" that highlight or privilege some aspects of what is represented, minimize (or even distort) others, and ignore or abstract away from a potentially unbounded amount of in-the-world detail.[9]

This commonplace is easiest to see in the case of conceptually structured representations—including not only propositional structures of the sort exemplified in logic (predications, conjunctions, disjunctions, negations, implications, quantifications, etc.), but also other representational forms that characterize the world in terms of a pre-established stock of objects, types, properties, and so on, such as those used in computer-aided design (CAD) systems, architectural blueprints, databases, and the like.[10]

---

9. Some philosophical readers may object that *names*, at least, refer to objects "without loss"—that is, to objects *as such*, and therefore do not impose even a trace of a conceptual frame. More later; for now I would just note that this would be true only if objects were objects (i.e., had determinate identity conditions) independent of being taken *as* objects, a standard realist position with which I disagree. See *On the Origin of Objects*.

10. Evans ties this notion of conceptuality to what he calls a "Generality Condition" (Gareth Evans, *Varieties of Reference*, Oxford: Oxford University Press, 1982, 100–105): a requirement that for a system to be able to entertain a "thought" or representation that a is f, it must also be able to entertain the thought that b is f, c is f, etc., for any terms b, c, etc. that represent other objects, and also that a is g, a is h, etc., for any predicates g, h, etc. representing other properties (modulo appropriate type restrictions). This is the sense of conceptuality that underwrites the characterization of the nonconceptual given in the sidebar on the next page; it differs from the one McDowell critiques in Lecture III of *Mind and World* (Cambridge, MA: Harvard University Press, 1996).

Nonconceptual Content

"Nonconceptual content" is a philosophical term used to explain the fact, evident upon reflection, that we are capable of thoughts and judgments that, although grounded in the ways of the world, and capable of being right or wrong, are not framed (in our minds) in terms of anything like a discrete set of articulable concepts. Classic examples include the speed at which one rides a bicycle, the position in one's egocentric personal space where items are located, shades of color, and so on. Adrian Cussins, an early theorist of the notion, when stopped for speeding on his motorcycle, famously answered a policeman's question about whether he knew how fast he was going with the statement "In some ways yes; in some ways no" (personal communication). Evans (MIT seminar 1978) reported on a footstep behind him when working in his study; in one sense he knew with exquisite precision where it was, yet would have been entirely unable to describe that location precisely in publicly shared concepts—feet, angle in degrees, and the like (see Gareth Evans, *Varieties of Reference*, 105). Similarly, it would be of limited help, if you were having trouble with your second tennis serve, to suggest that you hit the ball at the speed at which copies come out of your office copier.

But it also applies to most analog representations (those that represent continuous quantities in the represented domain—the domain of the problem to be solved, typically—with analogous real-valued quantities in the representational medium), since the property-to-property correspondence in such cases is typically assumed to be discrete at a higher-level of abstraction.[11]

---

11. As analog computers demonstrate, characterizing a situation in terms of discrete properties and relations does not mean that the *values* of those properties need to be discrete. The use of the differential calculus in classical physics is fully conceptual, on this account, even if any given velocity or acceleration or mass can be measured with a real

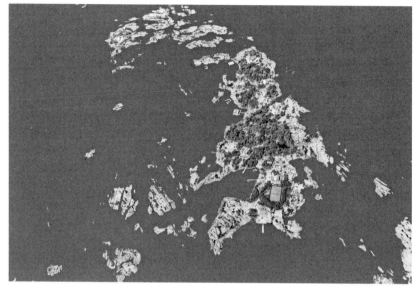

Figure 4[12]

The second point is also traditionally framed epistemo-logically. As noted in F4, theorists in numerous fields have emphasized the fact that not all human understanding has a "conceptual" form at all. The point is most familiar from phenomenological theorizing, but similar views under-write several contemporary currents in cognitive science, including enactivism, connectionism, and deep learning. Analogous intuitions have been explored in analytic phi-losophy in the notion of *nonconceptual content* (sidebar on the previous page). These approaches are united in viewing

...————————————

number. What matters is that the properties of velocity, mass, etc., are discrete. There is no such thing as something "halfway between a mass and charge," though even that might be expressible in terms of a higher-order discrete categorization (John Haugeland, "Analog and Analog," *Philosophical Topics* 12, no. 1, 1981, 213–225).
12. A masked version of figure 6, p. 34.

Figure 5

intelligence as emerging against an unconscious background—as I put it above, against an ineffable horizon of
tacit knowledge and sense-making.

Together with the perceptual points made earlier, these
views provide indirect support for an ontological world
view I have sketched elsewhere,[13] for which figures 4–6
provide a metaphor. Figure 4 (previous page) is a pho-
tograph of some islands in Ontario's Georgian Bay. Already, one can see that the real-world topography fails to
support the cut-and-dried ontological registration that
GOFAI assumed. Figure 5[14] "cleans the picture up" in a

---

13. *On the Origin of Objects.*
14. Figure 5 is by design simplistic; the filters used to create it are
not especially fine-grained. One could construct an image in which
they encoded vastly more detail—as, for example, is current practice in digital GIS systems. But the conceptual point would remain:

way reminiscent of GOFAI knowledge representation—making the islands, though still relatively detailed, "clear and distinct," and also internally homogenous, in the way in which conceptual models (such as data bases) all ultimately assume. While the question of "how many islands are there" may have a determinate answer in figure 5, the same is not true of the world depicted in figure 4. Distinctness flees, as realism increases. In the world itself, the question lacks a determinate answer.

More telling yet, though, is figure 6 (next page)—the same photographic image as figure 4, except now revealing the submarine topography. Compared to the world's messiness, the image is still simple: gravity is a single dimension of salience, the water line is relatively sharp, the image is gray scale, and so on. Nevertheless, if the islands in the image are taken as analogs for properties, then the images suggest what in fact is true: that as soon as one presses for detail, distinctions multiply without limit.

✦   ✦   ✦

...⎯⎯⎯⎯⎯⎯⎯⎯⎯⎯⎯⎯⎯⎯⎯⎯⎯

any (first-wave) knowledge representation structure is formulated in terms of a well-specified particular level of granularity; the representation "individuates" the world at a specific level of grain—no finer. In figure 5, as in all conceptual representations, there is a definite fact of the matter about how many islands there are, where exactly the boundaries are specified as being, etc. One would not learn more about the islands by studying the representation in any finer detail than necessary to reveal those facts. (This is what Dretske nonstandardly means by "digital" representation, as opposed to "analog"; see his *Knowledge and the Flow of Information*. Cambridge, MA: MIT Press 1981, 135–141.)

Someone might argue that the same is true of (even analog) photographs, because of the ultimate granularity of film. What is crucial, though, is *the same is not true of the reality that these images represent.* The world itself is arbitrarily detailed.

Paul Bennett Photography

Figure 6

Though our concepts and the properties they represent may seem discrete, in other words, the failures of GOFAI and the merit of the critiques mounted against it suggest that the idea of a "clear and distinct" world is an artifact of how it is represented. The issue for AI is that, in order to function in the world, AI systems need to be able to deal with reality *as it actually is*, not with the way that we think it is—not with the way that our thoughts or language represent it as being. And our growing experience with constructing synthetic systems and deploying them in the world gives us every reason to suppose that "beneath the level of the concepts"—beneath the level of the objects and properties that the conceptual representations represent—the world itself is permeated by arbitrarily much more thickly integrative connective detail. It is not just that our concepts sometimes have vague or unclear boundaries; it is that these facts tell on a world that itself is not itself clear cut. (Is a boisterous child the same as or different from a rambunctious child—or a boisterous CEO? If we are climbing the tallest peaks in Canada, and there is another local maximum 100 meters away from the summit we have just reached, do we need to go over there as well? Where does one "fog" end and another start? *Reality will not tell us*. If we want "clear and distinct" answers, we need to employ conceptual schemes that impose them.)

What are the consequences of these insights for AI? What follows from recognizing that the nature of reality is as suggested in figure 6: a plenum of surpassingly rich differentiation, which intelligent creatures ontologically "parse" in ways that suit their projects?

Some of the technical implications will be explored in chapter 5. Overall, though, it means that AI needs to take on board one of the deepest intellectual realizations of the last 50 years, joining fields as diverse as social construction, quantum mechanics, and psychological and anthropological studies of cultural diversity: that taking the world to consist of discrete intelligible mesoscale objects is an *achievement* of intelligence, not a premise on top of which intelligence runs. AI needs to *explain* objects, properties, and relations, and the ability of creatures to find the world intelligible in terms of them; it cannot assume them.[15]

How we **register** the world, as I put it[16]—find it ontologically intelligible in such a way as to support our projects

---

15. Ontology requires naturalization too, to put this in philosophical terms.
16. See *On the Origin of Objects*, and "Rehabilitating Representation" (discussed in chapter 2, page 15. note 18).

and practices—is in my judgment the most important task to which intelligence is devoted. As I will describe in more detail in subsequent chapters, developing appropriate registrations does not involve merely "taking in what arrives at our senses," but—no mean feat—developing a whole and integrated picture accountable to being in the world. It is not just a question of finding a registration scheme[17] that "fits" the world in ways locally appropriate to the project at hand, but of relentless attunement to the fact that registration schemes necessarily impose non-innocent idealizations—inscribe boundaries, establish identities, privilege some regularities[18] over others, ignore details, and in general impose idealizations and do an inevitable amount of violence to the sustaining underlying richness. This process of stewardship and accountability for registration, never imagined in the GOFAI project, is of the essence of intelligence.[19]

◆   ◆   ◆

One can (people do) say that GOFAI misconceived cognition, but as the foregoing discussion makes clear, I believe the deeper problem is that it misconceived the world. Why did it do so? One thing can be said, involving a subtlety that bedevils all discussion of AI systems' engagement with the world.

Two "views" or registrations of the world are relevant to

---

17. This process is sometimes described as identifying an appropriate "level of description" or "level of abstraction" (sometimes even as "coarse-graining"), but the metaphor of "levels" can be misleading.
18. That is: the metaphysical warrant for that which we register as a regularity.
19. The failure to imagine and therefore to take up these issues is a major reason, I believe, for GOFAI's inability to make progress on common sense. See the sidebar on the next page.

Common Sense

GOFAI's ontological presumptiveness, its blindness to the subtleties of registration, and its inadequate appreciation of the world's richness are the primary reason, in my judgment, for its dismal record on common sense. Notorious examples of early failure in this regard include systems suggesting boiling a kidney to cure an infection, and attributing reddish spots on a Chevrolet to a case of measles.[a]

The initial response to shortfalls of this sort, most famously taken up in the CYC project,[b] was two-pronged: first to codify, in a logical representation, all common knowledge to be found in an encyclopedia, and then, when that proved insufficient, to encode all common knowledge *not* in an encyclopedia. Those projects, and the wider "commonsense reasoning" program of first-wave AI, though still pursued in some quarters, have largely faded out of sight as models on which to base commonsense reasoners. They live on in such structures as Google's "Knowledge Graph," which is used to organize and provide short answers to queries in search engines. Tellingly, though, the results of such searches are snippets for interpretation by human intelligence, not the basis of the machine's own intelligence.[c]

---

a. Hamid Ekbia, *Artificial Dreams* (Cambridge: Cambridge University Press, 2008), 96–97.

b. See http://www.cyc.com, where this project lives on.

c. See my "The Owl and the Electric Encyclopaedia," *Artificial Intelligence* 47 (1991): 251–288.

---

the design or analysis of an AI system, or indeed of any intentional system at all, including people. The first is the designer's or theorist's registration of the world in which they are building the system, and into which they will deploy it, or in which they are analyzing a system or creature—that is, the world, as registered by the theorist or designer, that they expect that system to find intelligible, deal with, behave appropriately in. The second is the system's own registration of that world—the registration that will presumably lead it to behave and deal with the world in the way that it does. *There is no reason to suppose the two registrations will necessarily align.* As explored in greater detail in the sidebar on the following pages, the suggestion about why GOFAI may have misconceived the world is this: because first-wave AI and GOFAI took a techno-scientific attitude to their subject matter of constructing an intelligent system—analyzing it in terms of its causal or mechanical constituents formally conceived[20]—and because they did not deal with the phenomenon of registration at all, they may simply have assumed that intelligence could be constructed in a system that *itself* took a techno-scientific attitude to its world.

It did not work out that way.

---

20. For example, analyzing it in terms of what Sellars would call the "scientific" image, as opposed to the manifest image likely used by the system being modeled.

Theorist vs. Subject Registration

An issue that comes up when theorizing intentional systems is the relation between (i) the theorist's registration of the world toward which the system in question is oriented, and (ii) the registration of the world in terms of which that system finds that world intelligible. Take T, S, and W to be the theorist, system under investigation, and world or task domain, respectively. If T is a simple realist, taking W simply to be "is as it is," then the issue for T is straightforward: T can characterize W, or at least those aspects that S deals with, "correctly," and assume that S does—or should do so—as well.

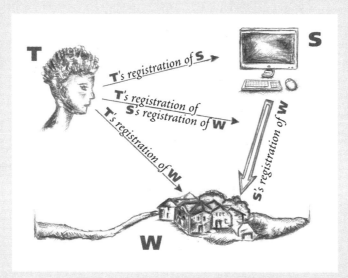

If registration is taken seriously, however—as we must in order to do justice not only to AI, but to the world and intentionality more generally—then at best it would be preemptively inscriptive, and at worst false, for T to presume that S's registration matches their own. T's registration may be adequate with respect to *nonsemantic* aspects of S's engagement—that is, adequate in terms of which to understand S's causal interactions with W (except in as much as those causal interactions need to mesh with S's intentional registration, such as serving as satisfaction conditions for S's planned actions). A much-touted feature of dynamical systems accounts of cognitive behavior, for example, is their ability to theorize the agent–world relation in the form of the differential equations required by dynamical systems theory.

When it comes to intentional or semantic relations between S and W, however, the situation grows more complex—as depicted in the figure on the previous page. Three (related) issues are at stake: (i) how T registers W, (ii) how S registers W, and (iii) *how T registers S's registration of W*.

Although this is not the place to engage in a detailed technical analysis of how these registrations can be meshed or appropriately coordinated, a few remarks can be made. First, to the extent that T's and S's registrations align, the issue is unlikely to be problematic. The problem grows more challenging to the extent that they differ—akin to that of understanding a foreign culture or alien species. There is no reason to suppose that T will necessarily be able to "grasp," at all, how it is that S registers W, no matter how committed T may be to doing so.

One might imagine that understanding human intelligence would be an instance of the former case, and therefore straightforward, but that is where the issue mentioned in the text comes to the fore. If T brings a scientific or technical theoretic attitude to bear on its theory of S, S's intentionality, how S perceives and thinks about the world, and so on, then T's registration of W is perforce likely to be *scientific*—treating W in terms of formal ontology: discrete objects, exemplifying clearly-defined properties and standing in unambiguous relations, and so on. Unless T is theorizing how S is a scientist, however, T's scientific registration of W and S's nonscientific registration are liable to part company.

These discussions illustrate one way to understand how first-wave AI and GOFAI may have "got the ontology wrong." It is easy to see how, given that GOFAI was blind to registration overall, it would have made an implicit presumption that elementary intelligence could be fashioned out of the GOFAI researcher's scientific or technical attitude toward W. Or put another way, GOFAI may have failed (ontologically) because it projected its own default *theoretic* attitude onto S.

I take it to be a consequence of the metaphysical position described in *On the Origin of Objects* that no deferential or accountable registration can be constructed from such a base.

4 — Transition

So GOFAI failed. Beginning around 1980, a variety of new approaches took hold. One, to be considered in the next chapter, shifted attention "downward," to consider the brain and brain-style computational architectures. This is the architectural approach that underlies AI's headlong embrace of machine learning and second-wave AI. It has developed in parallel with explosive growth in cognitive neuroscience, now arguably the most important sector of cognitive science. Together, these two projects currently have something of a grip on the intellectual imaginations of both AI and cognitive science.

Before second-wave AI took hold, however, four other "broadening" approaches gained prominence:

Embodied · Take the body seriously.
Embedded · Take context and surrounding situation seriously.
Extended · Maybe the mind is not just in the brain, or even brain plus body, but extends into the environment (which we, both individually and as societies, arrange and construct so that it can serve as a form of "cognitive scaffolding").
Enactive · Don't separate thinking from full-blooded participation and action.

The embodiment thesis not only involves consideration

of the whole body (limbs, torso, etc.) of the organism in question, and takes activity to be an integral component of intelligence and cognitive skill (for example in perception and navigation), but it also focuses on the physical properties and restrictions on the brain or processing unit (such as heat, energy usage, etc.).

These approaches were most vigorously defended during the 1980s and 1990s, in a period that has come to be known as the "AI winter"—a term descriptive of a lull in the AI funding environment, not per se a comment on the pace of intellectual developments, though not unrelated to the failure of GOFAI to reach its promises. The four theses were primarily advocated in cognitive science, but they also garnered some prominence in AI, where they continue to garner allegiance. They have not been embraced as such in the software engineering world, but if explicitly articulated might be endorsed by many members of that community—so long as the "body" of the computer were understood to be something like the CPU and directly affiliated memory, or at least the "local machine." Certainly their overall sense is manifested in the rapidly expanding networks of the world's growing computational infrastructure (e.g., in the widely trumpeted "internet of things").

I agree that human intelligence is embodied, embedded, (somewhat) extended, and (often) enactive—and that all four are critical considerations in cognitive science. But I think the reasons for GOFAI's failure go deeper than its failure to deal with these issues.

In brief, the reasons are three. The first has already been suggested: GOFAI's untenably rigid view of formal ontology. The second has to do with the semantic point made earlier (P4): that reference, representation, and semantic relations to the world are not effective, and therefore cannot be "seen" from a standpoint of blanket mechanism. The third arises from a failure to recognize the profound importance of taking our thoughts and representations and information to be *about the world*. We need to understand the conditions on world—what the world is, and what it is to hold something accountable to being in it, to being hosted by it.

5 — Machine Learning

Fast forward, then, to the present—to deep learning and affiliated machine learning (ML) technologies associated with second-wave AI.[1] These systems have made definite progress on the first and second of GOFAI's failures (neurological and perceptual)—and have arguably begun to address, though they have by no means yet fully embraced, the third or fourth (ontological and epistemological).

ML is essentially a suite of statistical techniques for:

1. the statistical classification and prediction of patterns
2. based on sample data (often quite a lot of it)
3. using an interconnected fabric of processors
4. arranged in multiple layers.

These techniques are implemented in architectures often known as "neural networks," because of their topological similarity to the way the brain is organized at the neural level. Figure 7 illustrates a way in which they are often depicted, but a better way to understand contemporary machine learning is in terms of the following four facts.

---

1. Though the phrase 'machine learning' was employed in era of first-wave AI, I will use it here, especially the 'ML' acronym, in its contemporary sense: to refer not only to deep learning algorithms but also to a variety of follow-on technologies, including deep reinforcement learning, convolutional neural networks, and other techniques involving statistical computations over complex graph configurations.

For deep learning in particular, see Yann LeCun, Yoshua Bengio, and Geoffrey Hinton, "Deep Learning," *Nature* 521, no. 7553 (2015): 436–444.

hidden layers

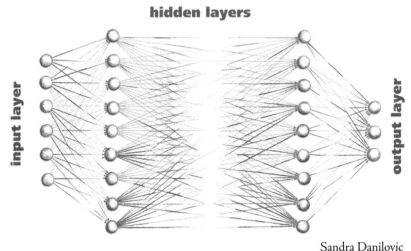

input layer

output layer

Sandra Danilovic

Figure 7

D1 **Correlations**: As we have seen, first-wave (GOFAI) systems were built to entertain and explore the consequences of symbolically articulated discrete propositions implemented as formal symbols representing objects, properties, and relations in terms of a presumptively given formal ontology. Based on this model, rationality and intelligence[2] were taken to involve deep, many-step inference, conducted by a serial process, consisting of one or a few threads, using modest amounts of information, formulated in terms of a small number of strongly correlated variables (sidebar, next page). Standard logical connectives, such as negation ($\neg$), conjunction ($\wedge$), disjunction ($\vee$), implication ($\supset$), and the like, procedure and class definitions, and so on, can be understood as various forms of 100% positive and

---

2. Or thought, or cognition—as indicated earlier, no distinctions were being drawn at the time.

negative correlation. The model makes sense under the classical assumption of formal ontology, particularly under the grip of Descartes's desideratum of "clear and distinct ideas."

As indicated in the sidebar, below, contemporary machine learning is essentially the opposite. It consists of shallow (few-step) inference conducted by a massively parallel process using massive amounts of information, involving a huge number of weakly correlated variables. Moreover, rather than "exploring the consequences" of such correlations, its strength is to learn and reproduce *mappings* between inputs and outputs. Whether the mappings should be understood as relating causal patterns in the machine

GOFAI vs. Machine Learning

The most compact way to understand the difference between GOFAI and machine learning is in terms of their opposing positions on five conceptual axes.

GOFAI

1. *Deep* (many-step) inference
2. By a *serial* process, using
3. *Modest* amounts of information
4. Involving a *relatively small* number of
5. *Strongly* correlated variables

Machine Learning

1. *Shallow* (few-step) inference
2. By a *massively parallel* process, using
3. *Massive* amounts of information
4. Involving a *very large number* of
5. *Weakly* correlated variables

(i.e., as uninterpreted mechanical patternings) or complex representations of configurations of the world (i.e., as interpreted) is a question we need to examine. Most literature appears to discuss it in terms of mechanical configurations, though the critical probabilities are always understood in terms of what is represented.

What is called "face recognition" is widely touted as an ML success. But like many other terms uncritically applied to computational systems, the term "recognition" rather oversells what is going on. A better characterization is to say that ML systems learn mappings between (i) images of faces and (ii) names or other information associated with the people that the faces are faces of. We humans often know the referents of the names, recognize that the picture is a picture of the person they name, and so forth, and so the systems can be used *by us* to "recognize" who the pictures are pictures of.[3]

To be cautious, I will mark with corner quotes ("⌜" and "⌝") terms we standardly apply to computers that I believe rely on our interpretation of the semantics of the action or structures, rather than anything that the system itself can be credited with understanding or owning. Thus: image or face ⌜recognition⌝, algorithmic ⌜decision making⌝, and so on. (Perhaps we should even say ⌜computing⌝ the sum of 7 and 13, but that is for another time.)

---

3. If the capacity is built into a camera, one might argue that at least the camera is computing a mapping between *real people* and other information about them. But the question of whether what the system associates with the other (represented) information is the person in view, or their representation on the camera's digital sensor, is vexed. See the discussion of adversarial examples in chapter 6, note 5 (p. 57).

D2 **Learning:** Perhaps the most significant property of ML systems is that they can be *trained*. Using Bayesian and other forms of statistical inference, they are capable of what I will call ⌐learning⌐—a holy grail of AI, with respect to which the classic first-wave model provided neither insight nor capacity.

Several architectural facts are critical to the capacity of ML systems to be trained. The complexity of the relatively low-level but extremely rich search spaces, architecturally manifested in high-dimensional real-valued vectors, enable them—given sufficient computational horsepower (see D4, below)—to use optimization and search strategies (especially hill-climbing) that would be defeated in low-dimensional spaces.[4] Equally important, at the relevant level of abstraction the correlation spaces need not be discretely chunked—allowing steady incremental transitions between and among states, the epistemological opposite of ideas remaining "clear and distinct."[5]

Metaphorically, we can think of these processes

---

4. The higher the dimensionality of the search space—the greater the number of independent variables—the less likely it is that hill-climbing algorithms (strategies that move in the direction that, locally, has the steepest upwards slope) will encounter local maxima.

5. It was never clear how Cartesian models could accommodate the gradual shifting of beliefs or of concept meanings, except by the excessively blunt addition or removal of specific discrete facts. Estimable efforts were made within the GOFAI assumptions, including the tradition of non-monotonic reasoning and belief revision or maintenance. See, for example, Jon Doyle, "A Truth Maintenance System," *Artificial Intelligence* 12, no. 3 (1979): 231–272; and Peter Gärdenfors, ed., *Belief Revision* (Cambridge: Cambridge University Press, 2003). But it is fair to say that learning remained an Achilles' heel of first-wave AI.

as moving around continuously in the submarine topography depicted in chapter 3's figure 6 (p. 34), making much less mysterious their ability to "come above water" in terms of what we linguistic observers take to be "discrete conceptual islands." That is not to say that the conceptual/nonconceptual boundary is sharp. Whether an outcropping warrants being called an island—whether it reaches "conceptual" height—is unlikely to have a determinate answer. In traditional philosophy such questions would be called *vague*, but I believe that label is almost completely inappropriate. Reality—both in the world and in these high-dimensional representations of it—is vastly richer and more detailed than can be "effably" captured in the idealized world of clear and distinct ideas. (There is nothing vague about the submarine topology of figure 6; it merely transcends ready conceptual description.)

D3 **Big Data**: Once trained, machine learning systems can respond to inputs of limited complexity (though often still substantial; a single image from a good digital camera uses megabytes of data). Training these systems, however, at least given the present state of the art, requires vastly more data. This is why machine learning is at present a "post Big Data" development; training involves algorithms that sort, sift, and segment massive amounts of it, culling statistical regularities out of an overwhelming amount of detail.[6]

---

6. Humans may require massive initial training sets, too—the idea being that early childhood may be a long training sequence for infants, in order to set up the initial prior probabilities needed for subsequent recognition and processing.

D4 **Computational Power**: Training algorithms can require phenomenal amounts of computational power.[7] Some systems in current use employ the parallel processing capacities of banks of GPUs (video cards)—up to thousands at a time, each capable of processing thousands of parallel threads at gigahertz speeds.

The last two points are historically significant. As Geoffrey Hinton has remarked,[8] they reflect the substantial truth in the (in)famous 1973 Lighthill Report,[9] which threw cold water on the idea that first-wave AI could ever scale up to produce genuine intelligence. Given not only the ideas on which it was founded, but also the amount of computational power available at the time, first-wave AI was indeed doomed. The million-dollar, room-filling computers on which GOFAI was developed[10] had less than a millionth the processing power of contemporary cellphones; banks of current-day parallel processing video cards can extend that power by yet additional factors of hundreds or thousands.

But AI moves forward. Using different ideas, masses of collected data, and radically improved hardware, the results of machine learning are genuinely impressive. Recurrent networks, deep reinforcement networks, and other architectures are being developed to deal with time, to push feedback from later stages in a process back to earlier ones, and so on. New accomplishments are published

---

7. This is especially true of those in use at the time of this writing.

8. Geoffrey Hinton, personal communication, 2018.

9. James Lighthill, "Artificial Intelligence: A General Survey" in *Artificial Intelligence: A Paper Symposium*, Science Research Council, 1973.

10. Primarily Digital Equipment Corporation PDP6s and PDP10s.

almost daily—transforming machine ⌜translation⌝,[II] ⌜reading⌝ X-rays, filling in deleted portions of images, and such. Certainly AI researchers are more excited and optimistic than they have been in 50 years; it is not just the press that is heady. I too agree that the developments portend profound changes to the nature of society and our self-understanding.

Does that mean that we have figured out what it is to think? I think not.

II. Google translation is especially impressive when the languages are similar linguistically and capable of similar registrations; increasingly less so as these similarities fall away.

6 — Assessment

How well do machine learning and second-wave AI deal with the four critiques of GOFAI?

6a · Neurological

The neurological critique may be sufficiently addressed. Inspired by what we know about low-level neural organization, contemporary machine learning architectures do to some extent mimic the brain. As suggested in chapter 3, though, the importance of that architectural similarity is not entirely clear. For one thing, it would be premature to assume that what matters for our brains epistemic power is our *general* neural configuration—shared by all higher mammals. That is all that current architectures mimic. Second, as programmers know, it is easy enough to implement one kind of architecture on top of another, albeit sometimes at a significant performance cost. While it is unlikely that evolution would have engaged in architectural mapping at the level of low-level neural circuits, the verdict is still out for the higher level forms of reasoning that are distinctive of our (so far) uniquely human capacities, which seem unlikely to be a simple consequence of general configuration. Third, there is the fact that no one knows whether working in the way our brains do is the only or even best route to general intelligence.[1]

---

1. If, as may turn out to be the case, the ontological structure of the world is such as effectively to require massively parallel networks for

Nevertheless, the parallelism of ML architectures, and perhaps their statistical abilities to deal with networks of probabilities, will likely be of lasting significance, perhaps particularly at the perceptual level (though it is ironic that ML systems currently take so much computational power to train, given that anything neurally realistic must be slow[2]).

6b · Perceptual

Regarding the perceptual critique, ML systems again seem closer to the mark. Their impressive performance on face ⌜recognition⌝ tasks, for example, is telling. As usual, moreover, the lesson is as much ontological as architectural. According to our best current understanding, it turns out that what makes faces distinctive are high numbers of complex weakly correlated variances across their expanse—the very sorts of feature that ML architectures are suited to exploit—rather than the presence of a few gross characteristics.[3] In general, visual ⌜recognition⌝ of scenes, and allied tasks such as ⌜reading⌝ X-rays, voice identification, and so on, are the types of task at which ML most evidently

···————————————————

interpretation, then that would be a reason for AIs to have such architectures—but the logical structure of such an outcome would be that AIs and brains would be similar for the same reason, not per se that AIs need to mimic brains.

2. See the discussion of Feldman's "100-step rule" in chapter 3, note 3 (pp. 23–24).

3. One might wonder whether there may not be a small number of characteristics in terms of which people can be recognized and identified, characteristics that we theorists have not yet discovered—or even that recognizing those characteristics is exactly what successful ML systems do. But if it takes a ML-type architecture to extract those characteristics from records of the physical light profiles reflected from faces, the point may be moot.

excels.[4] Successes in these realms are a substantial part of what has fueled contemporary excitement about the power of second-wave AI.

Still, it would again be premature to extract sweeping conclusions from these results on their own. It is sobering, to take just one example, that many of today's ML image recognition algorithms can be defeated by what to humans seem trivial changes in the images with which they are presented.[5] In the next chapter I will suggest one reason why this might be so.

6c · Ontological

Regarding ontology,[6] the issue is trickier (I will consider

---

4. *Pace* the standard caveat that calling such accomplishments "recognition" is a possibly culpable shorthand for saying that they are able to learn and repeat mappings between images of individual entities and some other computational structure associated with them.

5. For examples of such "adversarial" examples, see, for example, Athalye et al., "Synthesizing Robust Adversarial Examples," *Proceedings of the 35th International Conference on Machine Learning* (Stockholm, Sweden, *PMLR* 80, 2018).

As argued throughout the text, the fact that these adversarial examples work is evidence that current-generation systems may not in fact doing be *perception* at all, in the sense of dealing with an image of a distal situation—but rather something closer to mere image pattern matching, which we interpret as perception. That is, at best it is ⌜perception⌝.

6. As has been clear since the beginning, I use the term "ontology" in its classical sense of being the branch of metaphysics concerned with the nature of reality and being—that is, as a rough synonyms for "what there is in the world." (I defer discussion of the relation between ontics and onto*logy*, technically its *study*, to another occasion.) As in so many other cases, the term "ontology" has unfortunately come to be redefined, in contemporary computational contexts, to refer to structures that *represent* reality: classes, data structural types, concepts, etc.—licensing such otherwise inscrutable constructions as

the epistemological critique presently). Machine learning is not committed to any particular ontological story, so only indirect conclusions can be drawn. As already mentioned, moreover, discussions of second-wave AI typically focus on "uninterpreted" internal configurations (patterns of weights, activation strengths, transformations, etc.), obscuring whatever assumptions are being made about the nature of the world those configurations represent—about which different researchers, moreover, undoubtedly have different views. In addition, the relentless pace of ongoing research makes ML an unstable target of analysis. Still, it is not too early to say that the successes of second-wave AI provide evidential support both for the ontological critique itself and for the metaphysical view introduced in chapter 3.

It is in the realms of perception and action that machine learning most obviously overcomes the limits of formal ontology, and attends to the "subconceptual" terrain suggested in chapter 3's figure 6 (p. 34). When fed with data obtained directly from low-level sensors—visual pixels, haptic signals, and so on—ML systems have vastly improved on the levels achieved in GOFAI, even achieving levels competitive with human performance. This is enabled by a number of factors, including continuous patterns of weights, which allow for incremental adjustment and training, and sufficiently high dimensionality to "encode" all kinds of subtlety and nuance.[7] The resulting systems are particularly im-

---

"creating an ontology," and "ontological engineering." It is the world itself I am interested in here; I defer questions about its representation to considerations of epistemology.

7. It can be argued that the values of discrete pixels impose a "formal" grid on the array of impinging radiation, and therefore that the data are not genuinely continuous, but even if that were relevant a stream of values does not per se implicate an object, and the value of any such

pressive in not being defeated by intermediate cases, in accommodating noisy data, and in being robust in the face of ambiguity (as well, of course, as being able to be trained)—all of which capacities rely on the fact that they do not have to categorize and discretize their inputs at the outset.

In part, the accomplishments of contemporary systems stem not just from their being oriented toward a wealth of subconceptual detail, but also from their ability to store and work with it, rather than merely attending to it when initially presented—especially to integrate large amounts of information extracted from it into adjusted weights in the activation networks. This ability to ingest vast amounts of detail gives them a leg up on some perceptual tasks, and is critical in allowing them to move beyond what is humanly possible.

Our visual systems, too, seem capable of processing staggering amounts of low-level visual data when immediately presented with it, but it is less easy to imagine that we can store anything close to all of it, once effective coupling with the input is removed. Yet saying anything definite about human information retention is difficult, given our current ignorance about exactly how the brain works. Artists and visually-oriented people display stunning recognition ability for faces and individual scenes not recently encountered, for example—a facility that suggests not only nonconceptual but also informationally dense memories or predictive fabrics of expectation.

Still, in a point of considerable significance not only to AI but also to cognitive science and philosophy, it is commonly assumed that the role of perception, in the human case, is to take in the vast complexity of the perceptual

...——————————————————

readout is highly context sensitive, affected by incident illumination, the camera's position and orientation, and numerous other factors.

input and to output a "conceptual parse" of it—a conceptual parse of what is "out there," that is—arrayed in terms of familiar (and effable) ontological categories, no longer "burdened" by the wealth of detail that led to it. Once the perceptual input is categorized, that is, it is assumed, on the GOFAI model in particular and in many (especially analytic) philosophical models of mind, that an intelligent system can *discard the detail that led to that abstractive categorization*, and that reasoning or rationality from that point forward can operate purely in terms of the categories (i.e., purely in terms of sentences, propositions, or data structures categorically framed). This assumption fits into a general story that human categorization is at least in part a technique for avoiding information overload—abstraction in order not to swamp the capacity of the brain.[8] It is also, as we will see, the idea that underlies the Cartesian desire for "clear and distinct" ideas.

The success of second-wave AI suggests that reasoning need not work that way—and that it may not even work that way in humans.

One way to depart from the classical "discard the details" approach to classification is to avoid categories altogether. While we humans may classify other drivers as cautious, reckless, good, and impatient, for example, driverless cars may eschew discrete categories and chunking entirely, in favor of tracking the observed behavior of every single car

---

8. As technology advances, one might imagine that computational memory will be less easily swamped than our own. Time will tell, though it is sobering that even at today's state of the art, storing high-resolution video streams of all video cameras in operation remains challenging. Still, prospects of incredibly dense computational storage (e.g., DNA-based) will allow us to store many orders of magnitude more information that we do at present. How that will affect our environment and the fate of AI systems no one yet knows.

ever encountered, with that data then uploaded and shared online—participating in the collective development of a profile of every car and driver far in excess of anything humanly or conceptually graspable. Or to consider a different case, BlueDot, a Toronto startup,[9] collects worldwide travel itineraries, including billions of airline itineraries a year, to aid tracking and predicting the global spread of infectious diseases. Whereas traditional epidemiology rests on discrete categories or characteristics (middle-aged man, cancer sufferer, abuse survivor, etc.), no technical reason prevents a ML system from tracking all individual medical records, and dealing solely with vastly dimensioned vectors of real numbers, without any evident need to compartmentalize the data. The promise of "personalized" medicine, medical records with individuals' DNA sequences, and so on, may similarly "get in underneath the categories," to impressive effect.[10]

Even if a network does ⌜classify⌝ something, moreover—as a person, intersection, political dispute, war zone, whatever—it need not do so classically. Nothing in these architectures requires that, in "selecting" some conceptual category, the system must discard the trove of detail that

---

9. https://bluedot.global

10. There will be analytic challenges in how we understand such systems. Whereas a traditional diagnosis might be phrased as "you have a 52% chance of having melanoma" on a frequentist interpretation (i.e., because 52% of the people in this or that group that you are now identified with have developed melanoma), that may not be an appropriate way to cast a ML system's conclusion. It is not that probabilities will not pertain, just because a system is not dealing with groups; in fact most ML architectures are defined in terms of probabilities. The probabilistic diagnoses they come up with, however (or that we derive from their calculations) may require interpretation in something more like an epistemic measure of certainty: "I am 52% confident that *you in particular* have melanoma, based on what I know."

led to that result—detail that may provide information about the warrant for the classification, inflect it with ineffable shading and modulation, relate it to other concepts (neighboring islands), and so forth. In fact the very claim that the system *has* classified something may merely be a statement on our part, as external observers, that the patterns of weights and activations are "within the region" associated with the discrete labels "person", "war zone," and so on. Unless the system is required to make a hard-edged choice[11] among discrete alternatives, that is—such as to output a discrete token or word, corresponding to our human categories—even the distinction between whether or not a system has classified something need not be sharp.

Moreover, the success of ML systems in cases of simple reasoning shows that retaining and working with the statistical details and correlations derived from a "submarine" ontological perspective can convey substantial inferential power (making the reasoning for that reason at least partially nonconceptual). It is exactly such capacities that empower the widely touted era of Big Data. What is transformative about the present age is not just that we have access to mountains of conceptually represented facts, but that we have developed computer systems with predictive and analytic power enabled by their ability to track correlations and identify patterns in massive statistical detail, without having to force-fit those patterns of relation into a small number of conceptual forms.

---

11. Technically this should be ⌜choice⌝, but it would be pedantic to mark every possible instance of the distinction. Plus, we all employ what Dennett would call an "intentional stance" in our characterizations of computers (Dennett, *The Intentional Stance*, Cambridge, MA: MIT Press, 1987). I will mark just those cases where it is most important that we resist the tendency to attribute more capacity to the system than is warranted.

No one of these facts about the successes of second-wave AI is ontologically determinative; none provides invincible evidence of how the world is. But the more successful these systems grow, the more compelling the argument that the "coarse-graining" involved in interpreting the world through articulated concepts and discrete objects (i.e., interpreting it through the lens of formal ontology) is an information reduction strategy for purposes of calculation, reasoning, or verbal communication, rather than corresponding to any definite prior in-the-world discretization. Yes, we may talk as if the world were ontologically discrete; and yes, too, we may believe that we think that way. It seems increasingly likely, however, that such intuitions[12] reflect the discrete, combinatorial nature of language and articulation more than any underlying ontological facts, and also more than the patterns of tacit and intuitive thinking on which our articulations depends.[13]

---

12. A natural suggestion, from ML architectures, is that intuitions are (distributed) patterns of weights or activations formed in the high-dimensional networked representations of the richly interconnected submarine topologies of the world that underlie our concepts. The common difficulty we have "expressing" them may reflect the fact that words and discrete concepts are excessively bulky, insensitive tools with which to capture their ineffable modulation and subtlety.

13. These lessons are increasingly recognized within AI itself. Rich Sutton, a founder of computational reinforcement learning and leading ML scientist, recently put it this way: "We have to learn the bitter lesson that building in *how we think we think* does not work in the long run. ... [T]he actual contents of minds are tremendously, irredeemably complex; we should stop trying to find simple ways to think about the contents of minds, such as simple ways to think about space, objects, multiple agents, or symmetries. All these are part of the *arbitrary, intrinsically-complex, outside world.* ... [T]heir complexity is endless." (Rich Sutton, "The Bitter Lesson," http://www.incompleteideas.net/IncIdeas/BitterLesson.html, emphases added.)

In the face of second-wave AI, in sum, Descartes's idea that understanding must be grounded on "clear and distinct ideas" seems exactly backwards. The successes of ML architectures suggest that a vastly rich and likely ineffable web of statistical relatedness weaves the world together into an integrated "subconceptual" whole.[14] That is the world with which intelligence must come to grips.

✦   ✦   ✦

As always, care must be exercised when drawing such ontological lessons from the current state of the art.

First, we typically feed ML algorithms with data that are already processed, and to that extent "postconceptual": sex or gender selected from a short list of discrete possibilities, experience measured as various forms of unidimensional scalar, videos of traffic at what we humans classify as "intersections," even light intensity coming from some preclassified direction, and so on. Even if it looks on the surface as if the ML system is dealing with the world at a pre- or nonconceptual level, that is, there are many ways in which human conceptualization can sneak it—in ways unaccompanied by clarificatory subconceptual detail.

Questions need to be asked about the origin, appropriateness, bias, and so forth, of all such groupings and factorings—indeed, about the full range of data sets on which such systems are trained. If a ML system is given pixel-level detail about images of people or plants, it may not need to make a binary ⌜decision⌝ about whether some plant is a bush or a tree, or whether a person is brown or white. But if it is being trained on databases of images that have been categorically tagged by human observers, any subcategorical subtlety and traces of prejudicial nuance will in all

---

14. See my "The Nonconceptual World," unpublished manuscript.

likelihood have been lost, and the system is liable to fall, without ⌜knowing⌝ it, into derivative patterns of bias and prejudice. If fed data from Twitter, Facebook, and similar sources, for example, ML systems famously inherit and reproduce patterns of racism, public shaming, false news, and the like—all without, as it were, batting an eyelash.

We humans are of course affected by the discourses in which we participate, too, but one can at least hope that humans will bring a critical or skeptical attitude to such sources in a way that ML systems are as yet unable to do. Such reflective critical skills are exactly of the sort that I argue cannot arise from ever-more-sophisticated second-wave techniques. Instead, they will require what I will call full judgment.[15]

A second reason for caution in interpreting the success of second-wave AI stems from the fact ML systems are increasingly dedicated to sorting inputs into categories of manifest human origin and utility. To the extent that they are designed to mesh with our categories, even if they retain subconceptual detail, they will nevertheless thereby adopt, and be affected by, those categories' interests, utility, and bias. And if the outputs are discrete categorical

---

15. As mentioned in the introduction, it is possible that if second-wave AI were used as a basis for a class of synthetic creatures (perhaps along the lines of Sony's Aibos) that were themselves able to evolve, then over a long period of time such creatures might (as we did) eventually develop full-blooded rationality and judgment. See the discussion of "creatures" in chapter 10. The point is just that such a capacity would depend on their forming cultures and community, making and living by commitments, being governed by norms, going to bat for the truth, etc. They would not be capable of judgment, if indeed they ever reached that stage, merely in virtue of supervening on second-wave AI techniques. An *explanation* of their thereby developed normative capacities, therefore, would necessarily advert to more than their being just machine learning or second-wave systems.

classifications, as suggested above, and if we design the systems based on our classical myths about the nature of classification, the abundance of detail on which they rest, and thus any subtleties about the origins and appropriateness of such categorizations, are likely to be lost.

A third caution concerns a topic of some contemporary urgency: increasing calls for ML systems to "explain" their actions may prove to be curiously perverse. The abilities that the systems are being pressed to ⌐explain⌐ may be powerful exactly because they do *not* arise from the use of the very concepts in terms of which their users now want their actions accounted. The pressure to develop "self-explaining" or "interpretable" neural networks, that is, may inadvertently decrease their performance, and drive them toward unwarranted reliance on binary or discrete categories, toward implicit or even explicit reliance on formal ontology—may drive them, that is, back toward the epistemological and ontological inadequacies of GOFAI.

What should we make of all this? While the state of current research is messy, and clear-cut conclusions difficult to draw, I believe three ontological morals can be drawn.

1. The classical assumption of a discrete, object-based "formal" ontology is not a prerequisite of machine learning and other second-wave AI techniques. On the contrary, the success of ML systems, particularly on perceptual tasks, suggests a different picture: that the world is a plenum of unbelievable richness, and that the familiar ontological world of objects, properties, and relations, represented in articulated conceptual representations, is very likely "the world taken at a relatively high level of abstraction," rather than the way that the world is.

2. Much of ML's power stems from its ability to track correlations and make predictions "underneath" (i.e., in terms of vastly more detail than is captured in) the classificatory level in terms of which such high-level ontology and conceptual registration is framed.

3. The fact that ML systems are increasingly being targeted toward domains that have been ontologically prepared by—and targeted for—humans, typically in conceptually structured ways, inevitably leads these systems to inherit both the powers and limitations of human approaches, without any critical faculties in terms of which to question them. It is these factors that are giving rise to the widely discussed (but inappropriately described) phenomenon of "algorithmic bias."[16]

Years ago,[17] as noted in chapter 3, I outlined a picture of the world in which objects, properties, and other ontological furniture of the world were recognized as the results of registrational practices, rather than being the pregiven structure of the world. The picture is useful in terms of which to understand both the failures of GOFAI and the successes of machine learning. It depicts a world of stupefying detail and complexity, which epistemic agents *register*—find intelligible, conceptualize and categorize—in

---

16. It is the data—not only its form and its content, but also attendant factors about its selection, use, etc.—that is the primary locus of bias in machine learning results. The algorithms that run over the data are undoubtedly not innocent—requiring data sets to be formed in particular ways, etc. But most examples of bias cited in the press and literature are due more to skewed data than to culpable algorithm. We need critical assessments that properly tease apart the respective contributions of these two dimensions of ML architectures.

17. *On the Origin of Objects*, 1996.

order to be able to speak and think about it, act and conduct their projects, and so on. Most importantly, the view was developed from the ground up to take seriously the fact that most of the world—indeed, the world *as* world—outstrips the reach of effective access, necessitating disconnected, semantic representations that, for reasons of complexity, necessarily abstract away from most of its suffusing detail. It leads to a picture of conceptual prowess as most relevant to the intelligibility of relatively more distal situations, and nonconceptual skills as particularly appropriate for the suffusing detail of the immediately nearby. As I put it in another context:[18]

> "I sometimes think of objects, properties, and relations (i.e., conceptual, material ontology) as the long-distance trucks and interstate highway systems of intentional, normative life. They are undeniably essential to the overall integration of life's practices—critical, given finite resources, for us to integrate the vast and open-ended terrain of experience into a single, cohesive, objective world. But the cost of packaging up objects for portability and long-distance travel is that they are thereby insulated from the extraordinarily fine-grained richness of particular, indigenous life—insulated from the ineffable richness of the very lives they sustain."

Both the successes and limitations of first- and second-wave AI make eminent sense in terms of this picture—predictable characteristics of architectures beginning to extract the rich but radically simplifying registrations fundamental to perception and cognition.

---

18. "The Nonconceptual World," unpublished manuscript.

6d · Epistemological

What then about epistemology—subject matter of the remaining GOFAI critique?

Here the rubber finally meets the road. Two major issues, to be addressed in the next chapter, stand in the way of AI's reaching anything that can truly be called *thinking*. Both have to do with what is involved in holding thinking and intelligence accountable to the fabulously rich and messy world we inhabit. One, relatively straightforward, involves reconciling the first- and second-wave approaches—taking advantage of their respective strengths, and moving beyond at least some of their limitations. This integrative goal is starting to be recognized, and to be suggested as a necessary ingredient for third-wave AI.

The other challenge is more profound. I do not believe that any current techniques, including any yet envisaged as a subject matter of AI research, even recognize the importance of this second issue, let alone have any idea of what would be involved in addressing it. Explaining it will take us into realms of existential commitment and strategies for dealing with the world as world.

7 — Epistemological Challenges

Start with the first and easier goal, of integrating the merits of both first- and second-wave AI.

One of GOFAI's strengths was its ability to deal with **articulated reasoning**—inferences involving (sometimes long) chains of structured propositions involving implications, negatives, quantification, hypotheticals, and so on. "Since Randy and Pat are in Tokyo for the Paralympics, they won't be here for dinner on Saturday," for example. Or: "17% of London's population have parents who speak different mother tongues." Or even: "One reason why public support for higher education is proportionally lower in Canada than in the U.S. is that social healthcare, unassailable in Canada but not in the U.S., takes up a large fraction of the public purse, and Canadian voters are reluctant to devote much more money to other social programs—a category of which education is viewed as an instance." An illustrative list of the articulated capacities that even early AI systems were designed to deal with is given in the sidebar on the next page. They remain critical to general intelligence.[1]

---

1. The articulative capacities listed in the sidebar are interdependent, both syntactically and semantically. Some of the relations among them are theorized in cognitive science and philosophy under notions of *productivity* (the fact that cognitive production and comprehension are unbounded), *systematicity* (the fact that the meanings of whole sentences and whole thoughts are systematically related to the meanings of the words or tokens they are made up of), and *compositionality*

Characteristics of Articulated Reasoning

Some forms of conceptual structure that were common features of knowledge representation and models or reasoning in the GOFAI mold.

1. **Identity** and **nonidentity** ("Tully is Cicero," "the baker is not my aunt Hilda")
2. **Quantification** ("every Canadian owns a tuque," "there is a rattlesnake in the grass over there")
3. **Variables** ("marriages in which one parent of each partner immigrated from the same country")
4. **Logical operators** (and, not/negation, implies, etc.: "no one who is both a curler and a classical violinist likes garlic")
5. **Sets** ("the president, vice-president, and treasurer," "all bonobos in captivity")
6. **Opacity** and intensional contexts ("she said that France has a king," "he believes that π is rational")
7. **Categories** and **subcategories** ("angel investors," "everyone who was invited")
8. **Possibility** and **necessity** ("she might have transferred from Swarthmore")
9. **Default reasoning** ("unless otherwise noted, the tide must be considered in the design of all saltwater harbors")

...────────────────

(the fact that the meaning of a complex sentence or thought is determined by its grammatical structure and by the meanings of its constituents). The significance of these combinatoric relations for human cognition is bluntly suimmarized by Jerry Fodor, but the point would generalize to any computational system aiming at achieve general intelligence: "Human cognition exhibits a complex of closely related properties—including systematicity, productivity and compositionality—which a theory of cognitive architecture ignores at its peril. If you are stuck with a theory that denies that cognition has these properties, you are dead and gone." Jerry Fodor, "Connectionism and the Problem of Systematicity (Continued): Why Smolensky's Solution Still Doesn't Work," *Cognition* 62, no. 1 (1997): 109–119.

If GOFAI could deal with such logical intricacies, why did it fail? Because, as has been argued throughout, the ontology in terms of which it was formulated was all wrong. First-wave AI had no resources to ground its abstractions and conceptual symbolizations in the richness of the world, giving them a tendency to float free of reality—threatening them to devolve, as it were, to mix a little Derrida and Shakespeare, into an endless play of signifiers, signifying nothing.

Still, the internal structure of articulated reasoning is critically important, and must be a part of any genuine model of intelligence. Our current best understanding of it has been worked out in the context of logic, complete with its presupposition of formal ontology, clear and distinct concepts, and so on. Think about such paradigmatic logical expressions as "$\forall x[F(x) \supset G(x)]$" or "$[\varphi \supset \psi \equiv \psi \lor \neg\varphi]$." These formulae are normally interpreted with reference to determinate semantic and ontological facts about the meanings of their constituents. Any x is assumed to be clearly individuated and either F or not F, and G or not G, without ambiguity or matter of degree. Similarly for φ, ψ, and so on.

It was argued in the last chapter that perception, classification, nonconceptual reasoning, and even the sorts of reasoning employed in analyses of Big Data, do not require that degree of ontological clarity—in fact, that their power, as demonstrated in ML systems, arguably derives exactly from not assuming it. Per se, that does not address logical complexities of the present sort, though it is clear that human intelligence does not require perfectly "clean" categories in order to deploy commonsense versions of such logical relations. (The longer the chains of reasoning grow, the more likely conclusions will fall apart—grow false or unreliable—in proportion to the extent that the relevant objects and properties are not homogenous and sharp-edged.) Support for articulated forms of reasoning is one of conceptual classification's great powers—but how clean or "discrete" the categories must be in order for it to be usefully applicable is far from well understood. It is also clear that some degree of logical reasoning can apply to nonconceptual content ("that red sofa will not go well with the wallpaper in our living room," "write here on the board, not over there," etc.[2]).

These considerations raise a suggestion for AI, prefigured in the last chapter: to figure out what would be required to construct an architecture able to support ML's extremely rich, high-dimensional representational vectors as elements in complex patterns of articulated reasoning. Fuzzy logic might be considered an early stab in that direction, but it was restricted to single, real-valued truth values, and was "higher-order discrete"[3] in the sense that any given "fuzzy value" was exactly (rather than "roughly") what it was represented as being—that is, some exact real number. Rather, what is being suggested is to develop systems that integrate ML's ontologically rich and nonconceptually presumptive representations with the patterns of articulated reasoning paradigmatic of first-wave AI—not by gluing the two capacities together in a "bimodal" system, but seamlessly integrating them, so that the nuances, subtleties, adjustments, and so on, embedded in the underlying

2. These examples are considered nonconceptual because the sense of redness pointed to in the first is likely not redness in general, but the particular red of the sofa—a shade that the speaker is unlikely to possess adequate conceptual resources to describe, and because the patches of wall signified by "here" and "there," in the second, are unlikely to be well-defined regions with determinate boundaries.
3. John Haugeland, "Analog and Analog," *Philosophical Topics* 12, no. 1 (1981): 213–225.

subconceptual webs underlying the concepts being combined could play a role in establishing the subconceptual
webs underlying the result, in ways that would give nuance
and inflection to relatively immediate inference, but allow
increasingly long chains of articulated reasoning only the
more distilled, abstracted, and "discretized" the representations become.

The architectural implications of this proposal are non-
trivial. In the classical (GOFAI) case, complex hypotheticals, disjunctions, implications, and so on, can easily involve
dozens or even hundreds or more conceptually articulated
constituents. In current neural architectures, anything that
the systems ⌐know⌐ may be encoded in weights distributed
across the entire network; how to make states of the network into parameters or constituents of compositionally
structured other states of the network is neither evident
nor likely straightforward.

Nevertheless, this goal of compositionally structured
states backed by reams of subconceptual detail does not
seem fundamentally at odds with the sorts of problem that
AI researchers are already tackling—and indications that
the need for such projects is being recognized.[4]

---

4. See, for example, Gary Marcus, *The Algebraic Mind: Integrating
Connectionism and Cognitive Science* (Cambridge, MA: MIT Press,
2001); the papers in Joe Pater, "Generative Linguistics and Neural
Networks at 60: Foundation, Friction, and Fusion," plus comment
articles (*Language*, 95:1, 2019); and Hector Levesque, *Common Sense,
the Turing Test, and the Quest for Real AI: Reflections on Natural and
Artificial Intelligence* (Cambridge, MA: MIT Press, 2017).
  One challenge for such projects will be to determine whether helpful insights will emerge in the experience and techniques that have
been developed in contemporary machine learning systems trained on
the outputs of human conceptualization, such as Wikipedia articles,
Twitter feeds, etc. Along with the evident issues of bias, prejudice,
etc., discussed in the previous chapter, a serious issue that would need

♦   ♦   ♦

The other epistemological challenge is more profound.

No matter how otherwise impressive they may be, I believe that all existing AI systems, including contemporary second-wave systems, *do not know what they are talking about*. It is not that we humans cannot interpret their outputs as being about things that matter to us. But there is no reason to suppose, and considerable reason to doubt, that any system built to date, and any system we have any idea how to build, ⌜knows⌝ the difference between: (i) its own (proximal) state, including the states of its representations, inputs and outputs; and (ii) the external (distal) state of the world that we at least take its states, its representations and those inputs and outputs, to *represent*. And it is those external states of affairs that they are talking *about* (remember claim P2, about semantic interpretation).

What is required in order for a system to know what it is talking about? What is it that present day systems lack, such that they do not? That is the question to which the rest of the book slowly develops an answer. I can say now that it will at least require authenticity, deference, and engagement in the world in which that which is talked about

···————————————————————————

to be addressed, explored in the remainder of this book, is that the data sources these systems use as training sets are not held systematically accountable to the highly variable conceptions (registration schemes) in terms of which they have been formulated. Including the results of any such "data mining" without evaluating the registrational practice underlying each and every one of them (every article, every post) would constitute exactly the sort of "gluing together" that would defeat the aim of the project.

I suggested that this first integrative goal would be easier to address than the second, but at a deeper level doing it properly may depend on successful treatment of the second as well.

exists—and that neither interpretability nor "grounded interpretation" will suffice.[5] What this means will become clear, but some examples may help in developing motivating intuitions.

Suppose an automatic X-ray reading system constructs what we take to be, and therefore call, a 3D model of the lungs.[6] Call the model α. Does the system know what we know: the difference between α and *the lungs of which* α *is a model?* More generally, what is the computer likely to understand about the notion of a *model*, at all? Even if we encoded meta-level information into α stating that it is a model—that is, even if we were to add something like "MODEL(α)" to its data structures—how would the system know that that meant that α was a *model* of something in the outside world, in the way we do? Per se, meta-level information does not help; the problem simply recurses.[7]

By the same token, it is unlikely that AlphaGo and its successors[8] have any sense of the fact that Go is a *game*,

---

5. The engagement must be enough to secure the reference, which is something less than genuine engagement with that referent. I can refer to something that is outside of my light cone, even though engagement with the limits of light cone are proscribed. My ability to refer to a π meson, or to Mesopotamian culture, or to women's experience of sexism, are dependent on the intentional capacities of my culture and community; I cannot shoulder such intentional directedness on my own. But the "reach" to such referents cannot be so indirect, mediated, and externally brokered that reference is impossible, in the way that it is for Siri, which surely lacks the ability to refer to a pizza parlor in any substantive sense, if in fact it can refer at all.

6. A 3D model "under interpretation," needless to say—a model of the lungs' 3D structure, that is, not a model that is itself three-dimensional.

7. To know that "MODEL(α)" means that α is a model requires exactly what I am saying we as yet have no warrant in assuming—that the system is capable of reference and denotation (not merely ⌜reference⌝ and ⌜denotation⌝).

8. Including AlphaGo Zero.

with an illustrious millennial history, played by experts from around the world—or even, for that matter, and more pointedly, that there is a difference between the particular game it is playing and the representation of that game in its data structures. Siri and Alexa, similarly, though they may ⌜tell⌝ you about restaurants, bathrooms, thunderstorms, and operating system updates, do not really know what restaurants *are*, or bathrooms, or thunderstorms—likely not even system updates. As I have said since the beginning, these systems, being computational, are semantically interpreted, and so *we* understand their behavior as being about and referring to their represented worlds. But they do not understand *themselves* "under interpretation"—or realize that their thoughts and deliberations and utterances matter only under interpretation. That is why I question whether what they do warrants the label "understanding" at all. At best it is ⌜understanding⌝, but the stakes are high enough in these AI games that that might be a phrasing we do well to avoid entirely.

This is not to say that the structures and behaviors and ingredients of these systems are not interpretable by us; I take it that they are, as a condition on their being computational (P2). Indeed, that is exactly why Siri and Alexa are so useful. It is also not to say that contemporary computer systems do not act in the worlds that their representations are about. Many of them do, increasingly. Nor, importantly, is it to say that their symbols do not have determinate interpretations. This is not the place to rehearse what is wrong with Searle's Chinese Room argument, but let me just say that, in spite of some superficial similarities, I am not talking about formality, or about the legions of arguments (such as Searle's) that a formal system cannot have real semantics, because the interpretations are unanchored, and can be reassigned at whim. I would go to court to deny that the symbols in present day AI systems are "formal" in this sense—that is, in the way in which Searle understood that word. Some contemporary systems, in particular, are plugged into the world in such a way that their symbols are unambiguously grounded (think about transfer paths in internet routers, database entries in real-time financial accounting systems, email addresses, etc.). I would even argue that the semantic interpretation of many noncomputational symbols (e.g., words on signs and in books) are grounded by the practices in which they play a role. But neither signs nor books understand what their words mean, even if they have nonarbitrary interpretations. Rather, I am talking about something deeper—about whether the systems in which these (grounded) symbols play a role genuinely *understand* anything.

Put it this way. At the outset I said that semantics, in order to be semantics, must be *deferential*. At the moment, although we may design our systems with deferential semantics, *the deference is ours*, not theirs. If we are going to build a system that is itself genuinely intelligent, that knows what it is talking about, we have to build one that is *itself* deferential—that *itself* submits to the world it inhabits, and does not merely behave in ways that accord with our human deference. To do that, it will have to know (i) that there is a world, (ii) that its representations are about that world, and (iii) that it and its representations must defer to the world that they represent.

✦ ✦ ✦

What then is deference, what is the world, and what is it to know that there is a world to defer to in which objects exist? Those are the questions to which the rest of the book is addressed. But first something preliminary: Why does it matter? Why should we care whether our creations really are deferential—or, for that matter, whether they are genuinely intelligent? Would it not be enough if we could merely interpret them *as if they were* intelligent, deferential, and the rest?

No, it would not. If we are to trust their deliberations, AI systems need to be genuinely intelligent in order to be able to take responsibility for the adequacy of the abstractive ontologies in terms of which they register the world. Otherwise we should use them only in situations where we are prepared to take epistemic and ontological responsibility for every registration scheme, every inferential step, and every "piece of data" that they use along the way.

This is not a hypothetical problem. We are already presented with the results of data mining algorithms that have not just run over large individual data sets (such as census tallies) but surveyed vast collections of data sets, where we do not know what normative standards, registrations schemes, ethical stances, epistemological biases, social practices, and political interests have wrought their influence across the tapestry. Just as we would not trust an uneducated child or insensitive journalist to summarize the suicidal tendencies of teenagers at risk in diverse cultures, so too we should not trust an AI system unless we can similarly trust its ability to critically assess the merits, humanity, conceptual compatibility, legitimacy of assumptions, and so on, of all of the data sets that it surveys—to say nothing of those on which it has been trained.

8 — Objects

Some of the deepest thinking on these topics has been framed in terms of transcendental and existential philosophies. For insight we could delve into Kant's inquiry into the forms of sensibility and understanding as conditions of the possibility of knowledge of objects as objects. Or, turning to Heidegger, we could consider the existential question of the possibility of being toward the being of entities. But we do not need such fancy language here. We can ask a simple question:

> *What must be true of a system in order for it to register an object as an object in the world?*[1]

Again, this is not a question about what it would take for an AI system to represent (or deal with) something that we humans take to be an object in the world—or rather, to put the point more simply, just "register an object as an object," since, in line with the etymology, I will take *being an object* to mean *being part of what is objective*—that is, being part of the world. So the issue can be more compactly described as one of what it would be for an AI system to take something to be an object—for it to represent or refer to,

---

1. By "object," in this question, I mean nothing very specific. Except that it would lose the etymological connection with "objective," *entity* would do as well. That which is registered need not even be discrete or individual. At issue is what it is—what is required—for a system to *register anything, as such, in the world.* See *On the Origin of Objects.*

and thereby be deferentially oriented toward, something that *it* takes to be an object, that *it* takes to be in the world.

This is nothing that any computer system yet imagined can do. Nor do I believe that machine learning, nor any other second-wave AI technologies, nor anything I have seen proposed for third-wave AI, sheds light on it.

But that does not mean we cannot make progress. Several things can be said. Seven, in fact.[2]

Standards on Genuine Intelligence

I. ORIENTATION: The system must be *oriented toward that which it represents*, not merely oriented toward, or involved with, its *representation*. It must be "intentionally directed toward it," as philosophers would say. Referring to something is a way of being oriented toward it, so long as the reference is genuine—something I will explore. But reference is only one, particularly targeted, form of orientation. To be more general—to recognize that life is often mundane coping, navigation, and involvement in everyday projects, not theoretical reflection—we could use phenomenological terminology and say that the system must *comport itself* toward the object.

Surely, one might think, a computer can be oriented (or comport itself) toward a simple object, such as a USB stick. If I click a button that ⌜tells⌝ the computer to "copy the selected file to the USB

---

2. This list is not a theory of epistemology, or of existential commitment. In fact it is not a theory at all. The seven (nonindependent) standards are merely properties that understanding and full-blooded intelligence require—criteria that human adults meet, and that synthetic creatures should attain before we should dub them intelligent, and (more urgently) before we cede responsibility to them for tasks requiring genuine intelligence or judgment.

stick in slot A," and if in ordinary circumstances my so clicking causes the computer to do just that, can we not say that computer was oriented toward the stick?

No, we cannot. Suppose that, just before the command is obeyed, a trickster plucks out the original USB stick and inserts theirs. The problem is not just that the computer would copy the file onto their stick without knowing the difference; it is that it does not have the capacity to distinguish the two cases, has no resources with which to comprehend the situation *as* different—cannot, that is, distinguish the description "what is in the drive" from the particular object that, at a given instant, satisfies that description.[3]

It follows, in the present terminology, that the computer has no capacity to deal with the USB stick *as an object*. If, mid-copy, I rewire the write head, so that instead of writing to the drive, it sprays the bits onto my Facebook page, the computer would again be clueless, and not just in fact, but necessarily.[4] And because it would be congenitally clueless, there is no warrant for saying that it was ever oriented toward the *stick*, as opposed to (at most) being oriented toward the representation of the stick, or the driver mechanism that interacts with it.

How *could* a computer know the difference between the stick and a description it satisfies ("the

---

3. For philosophers: the computer is constitutionally unable to distinguish between *de dicto* and *de re* interpretations of (the computational analog of) "what is in the drive."
4. The point would remain even if there were a checksum calculation upon completion of the write; that too could be stealthily mimicked.

stick currently in the drive"), since at the moment of copying there need be no detectable physical difference in its proximal causal envelope between the two—and hence no way, at that moment, for the computer to detect the difference between the right stick and the wrong one? *That is exactly what (normatively governed) representation systems are for*: to hold systems accountable to, and via a vast network of social practices, to enable systems to behave appropriately toward, that which outstrips immediately causal coupling. To assume that a situation is exhausted by what is causally proximate at any given moment (to pledge a priori allegiance to blanket mechanism, that is) is exactly to be blind to representation, semantics, intentionality, and normativity—blind to being oriented toward the world.[5]

2. APPEARANCE VS. REALITY: What does it take to be oriented toward something? At a minimum, it means that the system must be able to distinguish the object from a representation of it—in order to be deferential toward the former and not toward the latter. That is, it has to be able to *distinguish appearance from reality*.[6] As already noted, it does not suffice to employ quotation or meta-level data

---

5. See the discussion of blanket mechanism in chapter 1 (pp. 3–4).
6. Some philosophers may feel that any distinction between appearance and reality requires a separation between the mind and a *mind-independent object* or a *mind-independent world*—contrary to the constructivist sensibilities endorsed throughout. I believe this way of framing the requirements for realism is vastly too strong. There is a huge and highly textured territory between indistinguishability and independence—a territory inhabited, in my view, by all of ontology, that is, by essentially everything.

structures in such a way that, as outsiders, we can take some of its representations (machinations, behavior, deliberations, whatever) to be about the world and others to be about its representations. The system must recognize (not just ⌜recognize⌝) that the object is *different* from its representation of it. To take an object as an object rather than an ⌜object⌝, that is—to refer to it *as* an object—requires "taking it to be out there." And since in general the object (targeted by its representations) will be beyond effective reach, the system must know that that toward which it is oriented will often be distal—outside the realm of effective connection, outside its immediate causal envelope (once again challenging the adequacy of causalist scientific accounts). Intelligence, that is, must achieve what I dub a "Robert Browning" criterion: *to know an object as an object is to know that it exceeds your grasp.*

Note again that the distal (noneffectively available) nature of semantic content so obviously applies to consciousness that we rarely notice how astonishing it is. What you have "in mind" (a friend, an impending exam, a truck coming round the corner) is not your interior, proximal mental states or processes that represent those phenomena, but the exterior, distal phenomena themselves. Moreover, because even objects to which you may seem causally connected, must, in order to be objects, have a past and future, both of which are beyond effective reach (physics prohibits direct causal connection with either past or future), even objects that are "present" transcend, as objects, that which is locally available.

We saw this divergence in the case of logic: a separation between how the system works (causally, mechanically) and what it is doing (semantically, intentionally, "under interpretation"). The present point is that, in order to be genuinely intelligent—in order, as I will soon say, to be capable of judgment—a system must (i) "know" the difference between the two, and (ii) through use of resources provided by the former, be oriented toward the latter.

3. STAKES: Not only must the system be able to distinguish appearance from reality; in order to refer to or be oriented toward an object, the system must *defer* to that object. To put it in Searle's phrasing,[7] the system must know that when "word" and "world" part company, the world wins.[8] Else truth would be sacrificed. In order to stand in a noneffective semantic relation to the distal world and to take there to be an object out there—to take an object *as an object*, for there to be reference at all—there must be stakes, norms, things that *matter*.

As Haugeland emphasizes, a system that is deferentially oriented toward the world will go to bat for its references being external, will be existentially committed and engaged. We need to know what this means, and to determine whether systems are, before we will be in a position to claim that any systems of our own devising are genuinely intelligent.

7. John Searle, *Speech Acts: An Essay in the Philosophy of Language* (Cambridge: Cambridge University Press, 1969).
8. Some may argue that one never has cognitive access to the world itself, only to representations of it, and therefore that there is no way for the world itself to win. I disagree with the premise, but even if it

4. LEGIBILITY: It is a precondition on a system's taking something to be something, distinguishing appearance or representation from reality, taking things to matter, that it find the object *intelligible*—or perhaps we could say *legible*—in the world.

What is it to be intelligible or legible? To be a patch or parcel of the world that is registerable within an accountable registrational scheme (more on this below)—that is, to be ontologically sound. Unpacking what that means does not require pledging allegiance to naive realism. We can avail ourselves not only of Kant but of Kuhn, Haugeland, social construction, cultural anthropology, and a spate of other contemporary resources. Ontologically, to *be* something—an electron, a faux pas, an on-ramp, a committee chair—is to participate in a constituted domain of regularities, rules, practices, configurations of reality in which we, as the epistemic knowers who are registering the world in whatever ways are appropriate to the constituted domain, participate, and to which we are committed. To be an electron is to fit into the whole physical weave of existence that electrons inhabit. To be a move or entity in a game, to use Haugeland's favorite domain (a knight fork,[9] a home base, etc.), can only be what it is within the context of the constituted regime of chess or baseball games. Even scientific or natural kinds, on Haugeland's (and perhaps Kuhn's) view, have roles to play in constituting regimes.

...———————————————

were true, the conclusion would not follow. The accessibility of the world does not bear on its status as normative arbiter of discrepancy.
9. A chess position in which a knight threatens two or more opposing pieces simultaneously.

5. ACTUALITY, POSSIBILITY, IMPOSSIBILITY: If there is to be a significant distinction between "getting an entity right" and failing to do so, there must be some feasible and nonarbitrary way of telling which is which, in particular cases. In fact it requires that the system embrace a three-way distinction, as regards any objects it takes to be objects:

a. What is the case about them
b. What is not the case about them, but *could* be the case (so that saying that such was the case would be false—an error, a mistake, something in need of correction)
c. What *could not* be the case about them—what is conceptually or ontologically impossible, so that a statement to that effect can be rejected outright, at pains of the whole intelligibility system collapsing (more on this below)

The system must be able to distinguish the actual, the possible, and the impossible—truth, falsehood, and impossibility.[10]

Research has given us insight into what exists, and has opened up our imaginations to what could be, in virtue of being founded on a set of laws which also dictate what cannot be the case, that rule out what is impossible. Protons cannot be grey, because protons are too small for color to apply. The number four was not seen on Queen St. this afternoon,

---

10. Haugeland has a more detailed account of this tripartite structure, including an analysis of what he calls the "excluded zone," in section 12 (p. 331) of "Truth and Rule-Following," in *Having Thought* (Cambridge, MA: Harvard University Press, 1998); see also "Truth and Finitude," in John Haugeland, *Dasein Disclosed* (Cambridge, MA: Harvard University Press, 2013).

because numbers are not occurrent. *Star Wars* movies may suggest that we might someday be able to zip across the universe in hyperdrive, but it is not going to happen.[11] No algorithm is ever going to examine all possible states of the chess board; no consistent formal system capable of ⌜stating⌝ its own consistency is going to ⌜prove⌝ its own consistency; and so on.

More pragmatically, some things that are not strictly speaking impossible we take to be impossible—such as the cup of coffee in front of me spontaneously leaping two inches up into the air (and simultaneously growing just enough cooler for energy to be preserved).

Why do we need the impossible, to say nothing of the false, in order for there to be import to things being correct—and thus for a system to be able to distinguish appearance and reality, and thus to be able to take something to be an object? Overall, it has to do with holding the whole system together, not only in order to distinguish ourselves and our representations from what we are thinking about and representing, but at the same time, to be explored in a moment, in order to hold the whole situation we are thinking about **to account**.

Holding things to account has a pragmatic purpose. If things present as impossible—if the evidence you are presented with, including that delivered by your sensory systems, suggests that something impossible has happened—you double

---

11. Long-distance quantum tunneling would not count as "zipping across the universe," in my book—but the point is that whatever does happen will have to accord with physical law.

down, check everything, reexamine your means of discovery, find alternative ways to discover the same entities, seek confirmation from other people, and so on. If I were so much as to begin to think that the cup of coffee in front of me leaped up two inches,

Multiple Registrations

All ways of registering the world are partial, skewed, appropriate in some circumstances and inappropriate in others. This foundational fact undergirds all knowing, reasoning, and intelligence. Unless a system can shoulder responsibility for acting in constant light of it, we should not trust its deliverances further than we can trust the adequacy of the registration schemes it employs.

The point has strong implications for any practice of integrating multiple data sources, including data mining and all other uses of "Big Data." Genuine intelligence requires making a seasoned judgment, whenever information is combined from different sources and circumstances, as to how the various data registered the world, how they can be soundly assessed, and what is required in order to integrate their different perspectives in a way that is accountable to the same underlying world. The issue applies at all levels, from the most far-flung collections of international databases to immediately adjacent posts in the same Twitter feed. Moreover, the integration task can never be fully delegated to a meta-level dictionary or translation scheme. Those are merely more registrations—partial, skewed, and contextually appropriate in their own ways. If a system is not itself capable of judgment, then it is we who bear responsibility for the differing perspectives and prejudices of all of its registrational sources, and for the legitimacy of all of its instances of data integration.

Contrary to current fashion, the mere ability to statistically combine data assembled from diverse sources without exercising judgment at each and every step may actually militate against, rather than for, anything we should want to call intelligence.

for example—if my perceptual system were to deliver that hypothesis to my cortex—I would not believe it, would not take the evidence as compelling. Instead I would conclude that I had blinked without realizing it, or that someone had jostled my desk, or that what I drank a moment ago was not coffee, or that an earthquake was underway, or something like that. That is, I would recognize that something "impossible" *seemed* to happen, but because impossible things do not happen—and this is the crucial fact—I will take that apparent impossibility as evidence that something has gone wrong, that some mistake has been made.[12]

In a way, this is no more than Kuhnian normal science. We operate within a registration scheme, hold objects and phenomena to account as being legible in its terms, take suggestions that they are not legible as evidence that something has gone wrong, dig in to repair the mistake, so as to be able to find out something that is correct—and thereby increase our knowledge. But in the present context, it is also relevant to why AI systems will have to be genuinely intelligent if we are to rely on them for general conclusions, because they have to be able not only to operate within registration schemes (ours or theirs), but to *hold those registration schemes to account throughout their use of them,* lest what they represent or register parts company with what is or could be the case.

---

12. This does not mean the world would not ultimately win, if this had in fact happened. Rather, the point is that it takes stronger and stronger evidence to adjust our sense of the world, proportional to the extent of how fundamentally our understanding must be revised.

Put it this way. There is no "right" ontology—no perfect registration scheme. As detailed in the sidebar on page 90, this fundamental fact has enormous implications for what it is to be intelligent, and for how we should view any system not capable of exercising judgment.

6. COMMITMENT: Not only must a system be able to distinguish appearance from reality—right from wrong—but it must care about the difference. This is a point on which Haugeland argued especially vehemently: doing the things we are talking about requires *commitment*. No creature—neither we nor the systems we build—can just incidentally come to know what is the case, can just happen to treat things as objects.

For an AI system to register an object as an object, that is, not only must there must be right and wrong *for it*, but that difference must matter, *to it*. Having systems do things that are right and wrong for us, having their actions matter to us—that is easy. That much is true of calculators, GPS devices, databases, and airplane guidance systems. Having things be right or wrong for us seems to be enough for route planning, and even for landing planes; it may be that it will be enough for driverless cars.[13] But it will not give the system itself objects, or a world, or intelligence.

For a system to care, its orientation to the world must be backed by a complex constitutive web of

---

13. It will surely be enough in sufficiently structured and contained environments. At issue is whether it will be enough to pilot cars in the midst of dense human activity. See the last paragraphs of section 6 of chapter 11 (pp. 126–127).

normative commitments. The system (knower) must be committed *to the known*, for starters. That is part of the deference: to take an object as an object, one must defer to the object, in order to sustain the appearance-reality distinction on which knowledge and intelligence, to say nothing of actuality, depend. But the commitments also put conditions on the knower. We must be such that we are committed to tracking things down, going to bat for what is right. Not only are we *beholden to the objects*, as Haugeland would say; we are also *bound by* the objects.

Explaining the requisite commitment draws us into existentialism. Being in the world, finding out what it is like, ensuring that what one thinks is in fact the case, and so on, requires existential commitment, without which the whole apparatus would dissipate, and one's thoughts or representations lose all of their significance—floating free of reality, like frictionless pucks in the void.[14] Commitment to the world, that is, constitutes the knower *as* a knower. The systems we currently use do not need that commitment, because we have it, and it is we who use them. But therein they fail to know, fail to be genuinely intelligent.

7. SELF: One more ingredient is necessary. An understander—human, AI, whatever—cannot take an object to be an object until that understander takes *itself* to be a knower that can take an object to be an object. That is: a certain form of "self awareness" is necessary in order to achieve the requisite detachment to be able to see an object as *other*, in order

---

14. John McDowell, *Mind and World*, 1996, 11.

for the system to hold *itself* accountable for being detached and holding the object to account. Again, the point is philosophically familiar, but typically phrased in daunting language (as Haugeland says: "Any disclosing is *at once* a disclosing of Dasein itself *and* a disclosing of the being of entities"[15]). But we can put it simply. In order for us to know about an object:

    a. It must be here in the world,
    b. We must be here in the world, and (recursively)
    c. We must know that both we and it are here.

It may seem surprising that the full substance of these seven points is not only something we profoundly depend on, but something that is required in order for us to identify the most mundane object as an object. But sure enough we do, and it is. Surprise merely indicates how deeply the ontological assumption underlying first-wave AI has been engrained in our consciousness—not just inscribed in our individual minds, but perhaps incorporated into the world view that undergirds contemporary technological society.

The failures of first-wave AI, and the fact that second-wave AI does not deal with them, should not be taken as undermining the magnitude or importance of the notion of an object in our lives—in our finding the world intelligible, in our being intelligent, in our ability to navigate, reason, and cope. On the contrary, it should heighten our awareness of the magnitude of the historical and sociocultural forces that have made objects so powerful and ubiquitous. That

---

15. John Haugeland, "Truth and Finitude," in *Dasein Disclosed* (Cambridge, MA: Harvard University Press, 2013), 190; emphases in original.

is not to suggest that the notion is innocent. As diverse cultures, poets, constructivists, and myriad others know, there are situations where objectification and reification are gravely problematic.[16] But for good or bad—most likely for both—taking the world to consist (perhaps among other things) of objects is a powerful ontological or registrational frame. No attempt to build synthetic intelligences can get very far without engaging with what it involves in its full substance and gravity.

With respect to AI, the success of machine learning has taught us that taking the world to consist of objects is not a necessary presupposition of synthetic or computational devices. More seriously, it has also shown that no genuinely intelligent system can *start* with an object ontology. If a system is to do justice to the world around it, if it is to understand what it is representing and talking about, it needs to be constructed or evolved in such a way as to *earn the ability to register the world in terms of objects*: objects grounded in the world, objects integrated into an unutterably rich metaphysical plenum—objects, as we will see, held relentlessly to account.

---

16. Needless to say, there are problematic particular objects in the world, too: nuclear weapons, lies, a few people some of the time. At stake here is the more general issue of what justice is done to the world to register a patch of it *as* an object.

9 — World

Underlying all the foregoing points—undergirding the conditions on genuine general intelligence—is something even more primordial. It has to do with that which I am calling the **world**.

In a way, the point is simple. Everything we believe, everything we take in, everything we represent and are committed to, must be something we can understand as being in a single world—the world that both we and it inhabit. Even if a phenomenon makes sense "on its own" (whatever that might mean), it cannot exist, cannot have the requisite actuality and otherness and so forth, unless it is part of the world—part of all that there is, that which is total, the "One." (See the sidebar on the next page.)

This generates a four-fold condition on our taking an object—or anything else—to be real:

Four-fold Commitment

1. We must hold the object accountable to being part of the world;
2. Reciprocally, we must hold the world accountable to hosting the object;
3. We must also hold ourselves, and our relationship to the object, accountable to being in that self-same world; and
4. Reciprocally, we must hold the world accountable to hosting *us*, and that relationship, as well.

If something appears to us that is unaccountable, something is profoundly wrong. We have to demur. We cannot go there. We must fight like hell to get out of that place, or we will die.

A few explanatory comments. First, honoring this standard does not require being *aware* of the four commitments, especially in any explicit sense. In the first instance, in fact, an explicit belief to the effect (especially a propositional representation of them) would not help. The four-fold commitment is prior; it must grip us as a governing norm, in order for any beliefs or states of awareness to be about their subject matters. Second, the "we" in the formulation does not refer to us as individuals, even if it ultimately places responsibility on individuals' shoulders. These are conditions that societies and cultures have hewn, over many centuries, into which we are indoctrinated as

Pluralism

Some will say that different people live in different worlds. I am sympathetic to pluralist ontologies (registration schemes, I would say, which configure ontologies). But there must be a "lower level" or, in some other sense, a more ultimate metaphysical unity holding everything together. If X shoots Y "in X's world," Y (that is, that which we or X register as Y) will likely die no matter what "world" they inhabit. If that which I register as a black hole destroys my world, it will likely destroy yours as well. If I want to reach out to you, not only would I be unable to reach you, if you genuinely lived in a different world; I would not even know of your existence. And so on. This is why reference (like a bullet) must "go on through" the registration scheme to reach the world itself. See section 12.a.

As I put it in *On the Origin of Objects*, doing justice to AI, to the human condition, and to the world requires embracing just the right combination of ontological pluralism and metaphysical monism.

children, in learning a language and taking our place as members of our social communities. Third, the commitments constitute a high standard. None of us meets them all at every instant of our lives, and we certainly need not be continuously conscious of them, but as I will presently argue, if we are adults we are accountable to them, and the civilized and cultural fabrics that enmesh us need to sustain the standards as parts of a civilized society.

To put this in cognitive science terms: to take an object to be an object does not just mean interacting with it, solving the "symbol grounding" problem in a local way, by associating uses of its name with that which it causally interacts with—the way an animal might cotton onto prey, or onto a rag doll, or the way an infant might attach to their mother. Though that kind of counterfactual-supporting interactive connection may be enough for some purposes,[1] it is not enough to constitute something as an object for the system in question,[2] and therefore not enough to warrant the

---

[1] Such as to support a Dretskian information link (see Fred Dretske, *Knowledge and the Flow of Information*, Cambridge, MA: MIT Press 1981).

[2] An animal will certainly recognize that which we register as an object in some sense—it is the "objecthood" that is demanding, of which I suspect animals are not capable. Strawson (*Individuals*, London: Methuen, 1959) describes a simpler form of registration in terms of *features*—roughly, property-like universals that do not require discrete individuated objects for their exemplification. A standard example is our interpretation of such phrases as "it's raining," which, on his account, takes the world to instantiate the feature "raining" without there needing to be any *object* that is raining. The suggestion is that pets may do something more of this sort—"It's Tiggering again." It is even possible that babies initially recognize their parents along similar lines: "Hurrah! More mama!" See Ruth Millikan, "A Common Structure for Concepts of Individuals, Stuffs, and Real Kinds: More Mama, More Milk, and More Mouse," *Behavioral and Brain*

system's being called intelligent. We observers might take such behavior as legitimating semantic interpretability; we might take that with which the system interacts to be an object. We might even say that the symbol is "grounded"— that it has determinate semantic interpretation, that being that symbol's interpretation is not a matter of mere whim. But throughout, in such cases, the whole pattern remains our registration—our object, our deference.

Three examples of failure in the four-fold commitment will illustrate. Many years ago, after a parapsychology talk at Duke's Rhine Research Center, the speaker asked a few of us, on our way out the door, why no one believed his results—given, he claimed, that his statistics were as good as any published paper in reputable psychology journals. "It is not that your statistics are bad," Güven Güzeldere observed, astutely. "They could be ten times better, but still no one would believe you. The problem is that *no one has any idea of how what you are saying could be true.*" The speaker's claims may have been individually coherent, but, as Güzeldere pointed out, overall they were unaccountable; they did not fit into, or conserve, our sense of the world being the world. And the world being the world is utterly essential. Sans that, all bets are off.[3]

··· ————————————

Sciences 21, no. 1 (1998): 55–65; and "Pushmi-pullyu Representations," *Philosophical Perspectives* 9 (1995): 185–200.

3. The point is not to bar hypotheses or observations that challenge deeply held worldviews. Some of the greatest scientific advances have come from just such cases (black-body radiation, for example). Rather, what troubled Güzeldere, and myself, was that the speaker was not prepared to shoulder responsibility for the fact that what he wanted us to believe ran counter to the entire world view on which we all relied. What the audience was keenly aware of, to which the speaker seemed oblivious, was the enormity of the epistemic and ontological burden that would need to be taken on—at the very least acknowl-

Dreams, to consider a second example, are to my way of thinking unaccountable in just this way. In fact I believe that it is their very unaccountability that makes it evident, from the outside, that they are dreams. I am in a room with two people; it morphs into an auditorium where someone is giving a talk; the podium they are leaning on is actually coffee cake, which I am eating as I ride my bicycle through the Pyrenees. Whatever! It makes no sense, but that does not matter. I do not go to bat to resolve the inconsistencies; my heart rate does not soar because my grip on reality has loosened.

A third example. Late one night, in graduate school, I was watching a movie in a shared house. The phone rang, startling me because it was about 1:00 a.m. I reached over and picked up the handset (this was the era of phones plugged into walls), whereupon, instantaneously so far as I could tell, the television went off, all the lights in the house went black, the entire neighborhood was plunged into darkness—and the phone kept ringing! In retrospect the whole thing was perfectly explicable;[4] relevant here is merely the intensity of my panic. My heart rate spiked because *my world seemed to break*. Though tiny, it was a terrifying crack. The threat of losing the world is mortal.[5]

◆   ◆   ◆

---

edged, and ultimately addressed—in order to make his case credible.

4. By coincidence there was a city-wide blackout at the instant that I picked up the phone. Also by chance, the handset I picked up, left by a departing roommate, was not plugged into the wall. The original and continued ringing came from another phone in the room.

5. A split second after the incident, after I realized that I was still in the living room, that the power had gone off, etc., I was overtaken by a much more mundane form of fear: of intruders, of the threat of violence, etc. But that was subsequent to the prior existential fright.

I said above that an integrated network of commitments undergirds our ability to take an object to be an object—to take something to be in the world, to be intelligent. It is not just that we are beholden to objects, and bound by them. We are also beholden to, and bound by, *the world as a whole.*[6]

What does it mean to be committed to the world in this sense? I have already said that to take an object to be part of reality we have to keep in mind what is true, what is not true, and what is impossible. Underlying that is a parallel set of norms and commitments about the world as a whole. It is not just objects and local phenomena that we have to register and find intelligible. Registering an object means finding it intelligible in terms of the rules and regularities that constitute the domain within which it derives its existence as the object that it is. But those rules and regularities and practices need to be accountable too. They must support, rather than undermining, the world's status *as world.*

Does that mean that we hang onto our constituting conceptual frameworks or registration schemes absolutely—because if we let go of them we die? Almost. We lean in that direction. We have to—in order to maintain our ability to distinguish appearance from reality. But we do not hang on to these constituting schemes tyrannically. We cannot—and we must not. Just as *objects* or *entities* are only found intelligible in terms of conceptual or registration schemes, so too are conceptual or registration schemes. The constituting rules and regularities underlying practices

---

6. That does not mean, at all, that the world is an *object*. It is not, and cannot be. If this were a different era, we might call it God—at least in Tillich's sense of "God" being a name for "the ground of being." The ground of being is roughly what I mean here by the word "world"—that to which I defer, that which is One, that which wins.

and regimes are only legitimate (that is, as the etymology betrays, "can only be read") *if they make sense of the world as world*. As well as holding objects accountable to constitutive regularities and norms, so too we must hold constitutive regularities and norms accountable to the world as the ground of being.

10 — Reckoning and Judgment

It is finally possible to explain the title of the book.

I have tried to sketch some of what is necessary for a system, human or machine, to be able to think about or be oriented toward the world beyond it—a world with which it copes, through which it navigates, in which it conducts its projects, about which it reasons, to which it is committed, to which it defers. The system must not only be embodied and embedded in this world; it must also recognize it *as* world. No less is required in order for it to distinguish appearance from reality, and choose reality; to recognize the difference between right and wrong—and choose right; to distinguish truth, falsity, and impossibility—and choose truth. No less is required in order for it to register entities, phenomena, people, and situations as such. It must register all these things in the world to which it holds itself, and all that it understands, accountable.

Such, to our inestimable benefit, is the legacy of the human epistemic achievement.

10a · Animals

I have not yet said anything about animals.

Nonhuman animals, I take it, are in one sense accountable to the world. There is a huge variety of kinds, needless to say; the forms of accountability pertinent to cockroaches will differ substantially from those applicable to jaguars, or to chimpanzees. Just how much variation there is—how much difference, how much similarity—I leave to others. And any such account must include the fact that we are animals too. No theory of humans and other animals can ignore the continuity of our constitution, evolution, and development.

How much we overlap with nonhuman animals (especially higher primates) on capacities that contribute to our intelligence is another subject on which I profess no special expertise. It is certainly true that nonhuman animals manifest species-appropriate forms of awareness. Some have exquisitely subtle and precise perceptual (or ⌜perceptual⌝) systems.[1] There are stakes for them; if they get things wrong, they may not survive, individually or as a species. They manifest their own forms of care, and are emotionally sensitive. As is increasingly remarked, in certain respects nonhuman animals outstrip us humans.

Nevertheless, I do not believe that nonhuman animals engage in the sorts of existentially committed practices of understanding of the world *as world* that I have been exploring here. It seems unlikely to the point of impossibility that they are capable of achieving the sorts of objectivity necessary in order to take an object as an object, and thereby to be truly intelligent in a human sense. We may say that a pet recognizes some well-loved object; but that is an object *for us*. Without the four-fold structure of normative commitment laid out above, which a pet seems unlikely to be bound by, it cannot be an object *for them*.[2]

---

1. Presumably backed by massively parallel neural networks.
2. Or maybe pets and creatures can and do register objects as objects, in at least a rudimentary sense. My point is not to defend any particular account of animals, but to get a sense of what is required in order to take there to be objects, and a world—a sense in terms of which we can then have a discussion of whether, and if so to what extent, various types of animal and machine can do that.

Two things can be said for nonhuman animals, though, even if they are incapable of registering an object as an object. First, they *participate* in the world—perhaps even existentially, at least in a rudimentary sense, each in whatever sense that makes sense for them. Second, and crucially, to the extent that they represent or register the world, they *do so in ways that mesh with how they engage, navigate, cope with—and are vulnerable to—it.* The world is presumably intelligible to them, even if not as world, exactly in as much as, and to the extent that, it matches their forms of life, stakes, biological needs, etc.

What matters about animals, that is—and for the present discussion as well—is the (evolutionarily hewn) fit between (i) how they take the world, or whatever patches of it they take, and (ii) how they conduct their lives, what they care about, and what they are vulnerable to. That makes something about their lives or even cognitive powers—if we want to use that phrase for nonhuman animals— *authentic.* I believe that much posthumanist writing overestimates their epistemic powers, but to the extent that animals do something like thinking and understanding, it would certainly seem that they do so in an authentic way.

I will use the term **creature** for systems, including animals, whose registrational powers do not extend beyond the ways in which they conduct their lives, what they care about, what they are vulnerable to, what they are existentially involved with. The term is useful because the development of AI-based "pets" (such as Sony's Aibo) suggests that we may be, or anyway may soon be, constructing computational creatures in this sense. As I have said a number of times already, it seems not impossible that synthetic creatures based on machine learning, active sensors and effectors, and so on, might not achieve something in the realm of the authenticity that I have said animals have evolved to possess. That is, nothing in this book argues that it may not be possible to construct genuine computational *creatures* (even if what it would be for constitutive norms to apply to synthetic pets, not merely for them to be at the mercy of our whims, remains obscure).

10b · Computers

The same is not true for computers in general. Most of the computational systems we construct—including the vast majority of AI systems, from GOFAI to machine learning—represent the world in ways that matter to us, not to them. It is because of that fact that we call them *computers*, or *information processors*, and also because of it that they have power in our lives, that they matter to us. What limits them is that, so far, nothing matters *to them*. To use a phrase of which Haugeland was fond: *they don't give a damn.*[3] Things will only matter to them when they develop committed and deferential existential engagement with the world (Dasein, perhaps, if we wanted to speak that way).

A possibly useful diagram of the difference is given in figure 8—a structure in terms of which to understand humans, nonhuman animals and other creatures, and most of the computers we have built. The forms of life that I have labeled "authentic" lie in the region of creatures, animal or machine, that register the world (or that we register as registering the world) in ways that are appropriate to, *and do not exceed*, the forms of their existential engagement in the world.[4]

---

3. Zed Adams and Jacob Browning, eds., *Giving a Damn: Essays in Dialogue with John Haugeland* (Cambridge, MA: MIT Press, 2016).
4. What is crucial to the creature's engagement is not the registrations themselves, but *the world being as the system registers it*, and the crea-

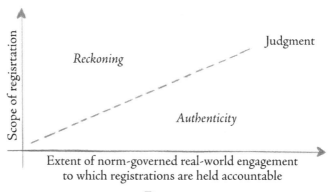

Figure 8

One way to sense the difference is to consider how different are our interactions with nonhuman animals from those with typical AI systems, the outputs of data mining services, and so on. Though we talk to animals, the fact that they do not use high-level language[5] keeps present before our minds the fact that their repertoire of feelings, understandings, projects, and the like, remain constrained to the realms in which they are able to participate authentically.[6] That is exactly not true when our iPhone ⌈tells⌉ us that there has been an accident on the highway, a medical

...———————————

ture's being existentially at stake to that fact.

I would not begin to suggest that this characterization could stand as a definition of authenticity. It is trivial to construct examples of systems that meet the criterion of being "under the line" in figure 12 that no one would count as authentic.

5. At least, so far as we know, they do not use compositional, productive, systematic languages (chapter 7, note 1, pp. 71–72).

6. If we register a nonhuman animal, such as a pet, in ways that do exceed that which is appropriate for how it engages with the world, and do so without intending to be metaphorical or ironic—such as when I say "my dog is frustrated with postindustrial society too"—then we are guilty of attributing a nonauthentic state to it.

system ⌜claims⌝ that nausea affects 32.7% of patients who take a given drug, or a wearable monitor ⌜suggests⌝ that if you have another drink you should not drive home. The fact that these systems ⌜use natural language⌝ can obscure to us the fact that they understand neither traffic nor nausea nor drunk driving. What is essential is that the increasing familiarity of such cases not blind us to the profound difference in types of system.

As noted in the introduction, I use the term "**reckoning**" for the representation manipulation and other forms of intentionally and semantically interpretable behavior carried out by systems that are not themselves capable, in the full-blooded senses that we have been discussing, of understanding what it is that those representations are about—that are not themselves capable of holding the content of their representations to account, that do not authentically engage with the world's being the way in which their representations represent it as being. Reckoning, that is, is a term for the calculative rationality of which present-day computers (including second-wave AIs) are capable.

I reserve the term "**judgment**," in contrast, for the sort of understanding I have been talking about—the understanding that is capable of taking objects to be objects, that knows the difference between appearance and reality, that is existentially committed to its own existence and to the integrity of the world as world, that is beholden to objects and bound by them, that defers, and all the rest.[7]

The terminology can be justified in various ways, beyond

---

7. It is because they do not exemplify this suite of properties, as I said in chapter 7 (p. 76), that current computer systems, including those constructed on the principles of second-wave AI, "do not know what they are talking about."

Hobbes's claim that "reason is nothing but reckoning"—a sentiment that many in AI believe, but with which it is clear that I disagree. But my aim is simple: merely to contrast reckoning with what we convey in plain English by saying that someone handled a situation with "good judgment."[8] By judgment I mean that which is missing when we say that someone *lacks* judgment, in the sense of not fully considering the consequences and failing to uphold the highest principles of justice and humanity and the like. Judgment is something like *phronesis*, that is, involving wisdom, prudence, even virtue.[9]

Why does judgment matter? The answer is again ontological. As I have been at pains to say, registrations (even nonconceptual ones) abstract from the world's detail—they approximate, do violence, privilege some things at the expense of others. That is not to deny that registrations

---

8. Two caveats. First, "judgment" is normally used in philosophy to signify an act, whereas I am using it here to name a property or capacity—something a system can *have*. Second, I am not here arguing that the notion of judgment being defended in this book is a "mark of the mental," in the sense of being a property that all human cognitive activity must manifest at all times, in order to count as human cognitive activity at all. It is important that we be able to say that someone did or said or thought something *without judgment*—by reflex, for example, or "thoughtlessly."

9. Phronesis traditionally concerns *practical* judgment, which is not my special concern here. Nevertheless, I take it that knowing how to act appropriately requires being grounded in concrete situations and recognizing that their subtlety and complexity may outstrip any registration of them—a theme I believe to characterize all judgment. In general, holding registration accountable requires resolute commitment to the world, not to registrations of the world—a theme that at least resonates with virtue ethics' deprecation of deontology (see, e.g., Alasdair MacIntyre, *After Virtue*, Notre Dame, IN: Notre Dame University Press, 1981).

are essential. Without them, given our finite brain (system) capacity and the world's partial disconnection, we would be at a loss, unable to orient to anything beyond our effective grasp. We would have no world.[10] Yet no matter their necessity, our *registrations* are not what matters—what matters is *that which we register.* To be accountable, to hold things to account, is to know the difference between the two—and to be committed to the latter, not the former.

Think of why we entrust child care to adults[11]—and about what can be catastrophically wrong with the response "But I did everything you said!" We want sitters to take care of our children, not of their (or our) registrations of our children. We want them to deal with any circumstances that arise, not just with those circumstances we have registered in advance. To reach out to the world in this way requires judgment, requires being an adult.[12] Or similarly: think of how pernicious it is to live with someone who is in love with their image (registration) of you, rather than with you "as you are." Or again: if a driverless car makes a mistake, what is vulnerable is the person represented in their data structures—the full person of ineffably surpassing richness, worth, and detail, far transcending anything the car could register or conceive. If we are going to drive cars that can shoulder commitments, that is what we want them to be committed to: the survival of ourselves and of others, not the execution of mechanical steps that their designers once thought would be the best calculative

---

10. That is, nothing would be world *for us*—nothing would exist, for us. "The world itself," as it were, would continue untroubled.

11. At least to emerging adults—and for that reason.

12. That is not to say that adults necessary succeed; or that how situations are registered is irrelevant to laws and justice. That is all secondary. What *matters* is the child; it is the child who might die.

route to that end. If they are incapable of such commitments, then they are mere reckoners, and we should understand and deploy them accordingly.

If representations and registrations came with God-given guarantees of sufficiency, judgment might not be so difficult, or so important. But they do not. And they cannot. And the world is not such that they could. Ultimately, you cannot deal with the world *as* world without knowing that—without judgment.

11 — Discussion

Seven comments.

1. HUMAN vs. MACHINE: As mentioned at the outset, the distinction between reckoning and judgment is not intended to get at, or to mark, any present or future distinction between people and machines.[1] For one thing, humans are likely also machines, in at least some sense, if the word "machine" is to have any merit (see sidebar on the next page). More seriously, I see no principled reason why systems capable of genuine judgment might not someday be synthesized—or anyway may not develop out of synthesized origins. It is not my purpose to outlaw the Golem—to banish synthetic constructions from obtaining the highest reaches of humanity.[2]

Also, as I have said repeatedly, by no means all human cognitive activity, especially in the small, meets the standards I am proposing for judgment. Perhaps the best we can say—though it is a lot to say—is that we hold human adults *accountable* for their actions. Even if they do not invoke full scale deliberative judgment in their every local move, we believe that they should act in such a way that they could do so if necessary. More generally, my purpose, in

---

1. Judgment is not a mark of the human. Perhaps it is a mark of the *humane*.
2. See previous note: the highest reaches of humānity, we might say.

framing judgment as a category, is to establish an appropriate standard on cognition such that it be-

Machines

The term "machine" has two connotations at least arguably inapplicable to people: being *constructed*, and being *mechanical*. Unless one is a dualist, however, the pertinence of the first to humans is difficult to deny unless one embraces, a priori, the belief that having children is not a form of "constructing" them. While that may seem obvious, it is not an easy thesis to defend. Needless to say, we do not understand "how children work" simply by participating in their natural procreation, but we increasingly do not understand the computer systems we devise, either—especially when their constitutive regularities (that which makes them the machines that they are) result from machine-learning algorithms, the details of whose "machinations" we do not grasp. If to be a machine requires being comprehended by engineers at the level of constitutive regularities, that is, then ML-based systems (image ⌜recognizers⌝, ⌜diagnosis⌝ systems, ⌜planners⌝, and the like) would have to be considered to be outside the realm of machines. (And understanding the regularities at just *some* level is not enough to separate us, since we largely understand how oxygen, hydrogen, carbon, and so on, work—out of which our organic selves are built.)

As for being mechanical, again unless one is a dualist, we are mechanical in some sense—just not *constitutively* mechanical (unless "mechanical" means some special way of being physical that we are not, but as above any such construal is likely to exclude computers as well). But then neither are computers constitutively mechanical; computation, I believe, is a constitutively semantical or intentional notion. So again, if people fail to be machines for that reason, so do computers.

Either we fall within the scope of the machinic, in other words, or, at least arguably, contemporary computers do not. Neither option supports a clear human vs. machine distinction relevant to AI.

comes an empirical question whether any given system (human or machine) can or does meet it.

Another way to look at this was suggested in the introduction. One of the challenges raised by developments in AI is how they will—and should—affect not only our sense of ourselves, but the constitutive standards on what it is to be human (since "human" is in part a normative predicate, not merely a biological one[3]). How should AI affect humanity itself, that is, not just the world that humans live in? Without raising judgment into too much of an ideal, we would hardly go awry if, by turning reckoning tasks over to our synthetic constructions, we could "up the ante" on people—raise the standards on what it is to be human—by establishing judgment of the sort described here as a defining bar on what constitutes authentic adult thought.

2. DETAILS: In suggesting a distinction between reckoning and judgment I have said little about what judgment comes to *in detail*. Myriad issues arise—such as whether the requisite values and existential commitments that judgment involves are or should

---

3. Alligators do not need to try to be alligators. Their claim on alligatorhood is secure, no matter how indolent, peculiar, or bone-crushingly violent. Humans are different: we must *strive* to be human. To commit an "inhuman" act is to fail to achieve the minimal ethical standards constitutive of membership in the fellowship of the human at least, perhaps even the "good." Homo sapiens is one thing; fully human, another. (Alligators need not worry about being "inalligator.")

The category "human," that is, is normative, not merely biological. It is affiliated with good and bad, better and worse—with values, morality, perhaps even spirit, those vexed interpretive issues, immortalized by the tree of knowledge, that Descartes clove away from the physical ones, back at the beginning of the Scientific Revolution.

be culturally local, generically human (humane?), or of wider compass.[4] I have suggested—and in fact endorse, but have not defended here—the idea that accountability to the world can serve as the grounds not just of truth but of ethics as well (accountability to the "One," to the preregistered, not to the postontological reality that most people take the word "world" to name). I have similarly left open such questions as whether the possession of judgment requires having a history, being enmeshed in a society, and so on. And I have said nothing about whether we can, or what it would be to, develop an amalgam of people, machines, institutions, practices, values, and the rest such that the assembly might be capable of judgment not attributable to the judgment of any of its members. And so forth.

Questions of this sort, I take it, should occupy our attention once we have established a workable understanding of the capacities of the systems we are building. My hope is that a first-order cartographic map of the territory occupied by people and AIs might free us from paralysis in the face of the overarching question of "whether intelligent machines are coming, and will take over." Once we have conceptual equipment adequate to the task of taking stock of them—of testing their mettle—we can take up the raft of detailed questions we will need to answer, even if only inchoately and provisionally at first, in order to reach our aim of sensibly deploying these systems in ways that are sound, beneficial, practicable, and sane.

---

4. See Alasdair MacIntyre, *Whose Justice? Which Rationality?* (London: Duckworth, 1988).

3. CONSCIOUSNESS: It will be argued that some of the characteristics I have attributed to judgment have more to do with consciousness than with intelligence—and will be claimed, for that reason, that it is unfair to pose them as criteria for AGI. There is something that is right about this comment, but much more that is profoundly wrong. Fundamentally, the objection gets the situation exactly backwards.

What is right is that some of the properties of full-blooded intelligence that I am describing—perhaps especially its dealing with a world of surpassing richness, exceeding the grasp of discrete concepts—are typically associated with "qualia" and other phenomenal characteristics recognized as being beyond the compass of discrete propositions and traditional logic. But to think, as a consequence, that the notion of judgment being defended here includes more within its scope than "mere rationality," and thus must be treading into the territory of consciousness (here) or emotion (next section), is to miss the fundamental ontological point that has underwritten the entire analysis.

If present-day theories of logic and of rationality, based on "clear and distinct" concepts, considered as given and unproblematic, do not provide an analysis of what it is to take responsibility for those concepts (responsibility for how they abstract, and idealize, and do both justice and injustice to, the worlds to which they apply), then *so much the worse for those theories of rationality*. Rich is how the world is, to which reason (not just feeling and consciousness) must be accountable. Various are the registration schemes suitable for characterizing it, including schemes beyond the reach of discrete concepts. That the world outstrips these schemes' purview is a blunt metaphysical fact about the world—critical to any conceptions of reason and rationality worth their salt. Even if phenomenological philosophy has been more acutely aware of this richness than has the analytic tradition, the richness itself is a fundamental characteristic of the underlying unity of the metaphysics, not a uniquely phenomenological or subjective fact.

More generally, I have tried to argue that making sense of first- and second-wave AI, on the one hand, and of the difference between reckoning and judgment, on the other—two aims of this book—requires rejecting a widely accepted divide between: (i) a cut-and-dried realm of conceptually discrete calculative rationality, of the sort that GOFAI tried to emulate, and that has perhaps unduly influenced Western metaphysics and epistemology for too long, based on a flawed ontological presupposition of a world of unproblematic discrete objects and properties; and (ii) a qualitatively richer, more fluid realm of "subjective" consciousness attuned to the subtleties and nuances of the world's resplendent underlying richness.[5] Second-wave reckoning (not even judgment) is succeeding by recognizing the world itself as ineffably dense. Judgment requires holding rationality relentlessly to account for this

---

5. It is "underlying" only with reference to the short-sightedness of an excessive allegiance to conceptual abstraction. As machine learning should if nothing else alert us to, there is nothing "hidden" about the world's indelible variety.

abundance, constantly ready to relinquish allegiance to registration schemes when they do it too much injustice. None of these things have uniquely to do with the first-person subjective character of consciousness.[6]

Put it this way: logic and GOFAI failed by viewing rationality as a postregistration—"post-ontological"—process of articulated moves, on the model of a chess game. Machine learning is helping to open our eyes to what phenomenologists have long understood: that holding registrations accountable to the actual (nonregistered) world is part and parcel of intelligence, rationality, and thought. If these insights lead to a reworked understanding of rationality, reason, and thought, which in turn impinges on and adjusts our understanding of consciousness (as I believe it should and will), well and good. But that is the direction of the implication: from the nature of reality to consciousness, not the other way around.

4. EMOTION: The distinction between reckoning and judgment is also not a difference between reason and emotion. Some may feel that being committed, going to bat, giving a damn, and so on, must be emotional states, in virtue of being action-inducing and motivational. For many reasons, I believe this

---

6. Consciousness does have its peculiarities, including its ineliminably first-person, unshareable character, which I will address elsewhere (e.g., in "Who's on Third? The Physical Bases of Consciousness," unpublished manuscript). I believe, however, that the identification of phenomenal qualia as uniquely characteristic of consciousness and as scientifically inexplicable are both mils, deriving from an ontologically inadequate understanding of the nature of reality.

is radically incorrect. Authentic judgment requires *detachment* and *dispassion*, for starters[7]—freedom from some of the very vicissitudes most character-istic of emotional states ("set your emotions aside," we appropriately tell children, when guiding them toward considered judgment). Interpersonal com-mitments, too, must similarly reach well beyond emotion, in my book, if they are to endure.

But the present point is not to impugn emotion, which some will argue to be essential to achiev-ing humanity, instilling compassion, and so forth. Rather, as suggested above, I want to reject the idea that intelligence and rationality are adequate-ly modeled by something like formal logic, of the sort at which computers currently excel. That is: I reject any standard divide between "reason" as hav-ing no commitment, dedication, and robust engage-ment with the world, on the one hand, and emo-tion and affect as being the only locus of such "pro" action-oriented attitudes, on the other.[8] It would be ruinous, I believe, to phrase this in terms I owe to Charis Thompson,[9] to cede *logos* to reckoning, and

---

7. See the introduction, note 2 (p. xv).

8. Among other issues, if one draws into the same category of emotion not only dispassionate concern with the truth of one's thoughts, the adequacy of one's registration, and the detachment of one's perspec-tive, but also self-indulgence, bias, prejudice, self-interest, revenge, and so on, one would then require the wherewithal with which to distinguish "good" emotions, as it were, from "bad" (the former from the latter)—a discriminating task, in my book, that would need to rely on dispassionate judgment.

9. Charis Thompson, personal communication, 2018. See the intro-duction to her *Getting Ahead: Minds, Bodies, and Emotion in an Age of Automation and Selection* (forthcoming).

force judgment to take refuge in *pathos*. Even the increasingly popular suggestion that "good thinking" requires combining rational understanding with emotion and affect buys into this impoverished conception of rationality.

As an example of such a misconception, consider Thomas Friedman's suggestion[10] that since industrial era machines replaced work of human *hands* (physical labor), and second-wave AI systems are now taking over the work of human *heads* (mental labor), what remains for humanity is labor of the *heart*: passion, character and collaborative spirit. Though superficially elegant, the suggestion is fatally flawed. It "disappears" the deliberative judgment that I am arguing is fundamental to reason itself, and is thereby blind to anything approaching genuine intelligence or dispassionate inquiry. Determining, standing for, and going to bat for the truth—all foundational components of reason—require active and "resolute commitment," as Haugeland puts it, far beyond anything computers can yet do.

5. RESPONSIBILITY: Fifth, only a system capable of authentic judgment can truly shoulder *responsibility* for its actions and deliberations. If an automated guidance system causes an airplane to miss the runway and to slam into a mountain, it is unexceptional to say that the system *failed*; we might even say that it was *at fault* (though I would want to write as ⌜*at fault*⌝). But we would—and should—balk at holding any presently constructable computer system *responsible* for the tragedy. An ethical stance can be

---

10. "From Hands to Heads to Hearts," *New York Times*, Jan. 4, 2017.

shouldered only by an existentially committed system capable of judgment.

Historically, it has been said that responsibility
can be shouldered only by a system capable of *human* judgment. One of my aims in eschewing the
human/machine distinction is to make room for the
development of a notion of judgment substantial
enough for the qualifier "human" to be neither necessary nor sufficient. We need to be able to ask where
responsibility for a given task is and should be located, without prejudice as to whether that judgment
resides in person or machine (or in policy, practice,
law, community, etc., or an amalgam of them all).[11]

6. ETHICS: For two reasons, I have intentionally not
taken up the topic of ethics. It has long been my
view, first, that substantial discussion of ethics and
AI requires a deeper understanding than we have
had, to date, of the issues I have been addressing
here: what AI is, what intelligence requires, and so
on. This book can thus be viewed as an attempt to
lay out some of the understandings that I take to
be prerequisite to meaningful conversations about
the ethics of AI. Second, as this book again demon-
strates, I do not believe, although it seems tempting
to some, that approaching the subject of AI through
an ethical lens is the best route into the deepest and
most serious issues about its nature, capacities, role,
impact, and proper deployment.

---

11. This injunction must be treated with care. It would be disastrous
to water down what responsibility is, what it takes, and the standards
to which we should hold both people and machines accountable, if
they are granted it, under pressures to delegate judgment to first- and
second-wave AI. See pages xix–xx of the introduction.

One preliminary ethical comment can be made, though, preparatory to any substantial discussion. On the one hand, many of the (critical) ethical questions raised by the development of reckoning technologies are unprecedented in detail, but of a kind familiar from the deployment of any new technology: genetically modified organisms, social media, nuclear weapons, and so on. Entirely different considerations arise, unprecedented except perhaps in the context of child rearing, when we ask what it would be for AI systems *themselves* to be moral agents—that is, to be able (and hence mandated) to take ethical responsibility for their own actions. I believe we are a long way from this, but the question will arise if and as our creations start to "think for themselves." My view is straightforward: such systems must be capable of *moral judgment*, understood as a species of exactly the sort of judgment under discussion here.

If does not follow that an AI capable of being a moral agent must for that reason be capable of employing ethical concepts *explicitly*, in the sense of holding or reasoning with an explicitly articulated moral theory. Just as a young adult might demonstrate estimable ethical judgment and behavior (perhaps even preternatural wisdom) in the absence of an explicitly held moral theory, so too might a synthetic system. Contrapositively, it is easy to imagine a reckoning system devised to calculate and manipulate symbolic representations of ethical worths and values, which did not even have the ability to refer authentically to the entities we would take its intentional states to represent—a system, that is, inherently incapable of ethical behavior at all. What grounds ethics, in person or machine, is the moral character of the subject's (semi-autonomous) judgment, intentional orientation to, deference, and engagement with the world—not, in the first instance, its possession or lack thereof of an articulated (ethical) theory.

A postscript might be added on a topic of intense current debate: the issue of driverless cars. To clear a potential distraction aside first, one of the difficulties of raising "trolley problems" in this context,[12] perhaps reflected in the fact that humans rarely know what they would do in such situations, is the presupposition that driving requires deliberative ratiocination. The human case is complex; we build roads and constrain traffic patterns and develop complex social practices so as to minimize the cognitive demands on human drivers; it would actually be dangerous for human drivers to need to assess, anew at every instant, whether the road continues around the corner or drops off a cliff, whether the vehicle in front will be grabbed by a hook rising from the road's surface and instaneously come to a halt, etc. Human drivers are held accountable only within highly regimented domains in which

---

12. The original trolley problem, popularized by Philippa Foot but originating at the beginning of the twentieth century, concerned the difference between doing nothing or intervening to affect the outcome of a situation with varying moral consequences. In discussions of driverless cars, the problem is usually simplified to one of merely choosing between two courses of action with distinct moral consequences (such as killing more elderly people or fewer children, or giving higher priority to passengers than to pedestrians).

as much judgment as possible has been shouldered and prefigured by material arrangements, social practices, and norms.

That said, the issue we face as a society is one of configuring traffic in such a way as to ensure that vehicles piloted by (perhaps exceptionally acute) reckoning systems can be maximally safe, overall. It may be that in some situations—long distance highways, for example—we can restrict the driving context sufficiently, as we currently do for aircraft, so that reliable reckoning confers adequate safety. That is, one possibly sane way to deploy currently imagined driverless cars would be to ensure that the various kinds of information available to them can be combined, *via reckoning alone*, to lead to outcomes we humans judge to be safe. Then, if and as we are able to develop systems that approach anything like judgment, the contexts in which they could be safely deployed will proportionally increase. But I do not believe that automated judgment, as I have emphasized throughout, is on any immediate horizon.

The conclusion generalizes. We should not delegate to reckoning systems, nor trust them with, tasks that require full fledged judgment—should not inadvertently use or rely on systems that, on the one hand, would need to have judgment in order to function properly or reliably, but that, on the other hand, utterly lack any such intellectual capacity.

7. CONSTRUCTION: Finally, I have said nothing about how judgment might be constructed. While in detail I do not believe any of us know, a number of points can already be made.

Note, for starters, three facts already stated about judgment's constitution. First, all intentional and semantic properties, of which judgment is an instance (along with being right, telling the truth, being reliable, etc.) are noneffective, relational properties.[13] They are not facts about the mechanical (effective) shape of what is in the machine.[14] So any question about the presence of judgment is a question about the system "under interpretation," not about its mechanical form. Second, as has also been emphasized, judgment is a holistic property of how a system works—how things are tied to other things, what the system does when encountering conflicting evidence, what other events or processes one perceptual event will trigger, how it responds to situations overall, etc. Even under interpretation, therefore, judgment is unlikely to be an isolable or categorical property of an architecture. Third, judgment involves a system's giving priority to the world, over its internal states and representations—and over the patterns exemplified by any inputs, machinations, and "interior" processes. What "giving priority to the world" looks like, from an internal architectural point of view, is at this point anyone's guess.

---

13. Calling them relational properties is technically suspect, because in general a relational property would be taken to be one that related two (or more) other ontologically secure entities. On the metaphysical and ontological world view being suggested here, however, registrational properties are partially (not wholly!) constitutive of anything's being the thing that it is, and so the entities "related" by the intentional or semantic properties are not necessarily ontologically individuable except in the relational properties' terms.
14. See my forthcoming *Computational Reflections* for an account of how even internal reference, where the target of the intentional act is an interior process or structure, is not an effective relationship.

Several consequences follow directly. First, the presence of judgment within a system is unlikely to be detectable at any level of implementation below the full system or personal level. It is unlikely ever to be realized by a structure or pattern of activity *identifiable as such* within the system's internal architectural configuration. Even less is it likely to have an effectively detectable neural or mechanical correlate (just as the presence or shape of *true* sentences within a logical system cannot be determined purely on the basis of syntactic shape). No fMRI of a person, that is, and no analogous description or diagram of the effective state of a synthetic system, is ever likely to reveal whether the system has anything approaching reliable judgment.[15]

Second, because of the constitutive importance of giving priority to the world over any internal representations or states, judgment is unlikely to be achievable by any system that does not engage with the world itself, in ways appropriate to what is required to hold its judgments to account. Judgment is available to systems and organisms that participate in the world; it will never be accessible to a mere bystander or disconnected reckoner.

Third, as already remarked, I do not believe that genuine intelligence can be explained within the bounds of science as currently conceived—from within a methodology of blanket mechanism, that

---

15. Someone might argue that there may be neural or mechanical *correlates* of judgment rather along the lines of the presence of a chemical or hormone corresponding to existential anxiety, which would perhaps not exist in a purely reckoning system—but these are speculations of fantasy rather than anything rationally defensible at present.

is—because of (at least current) science's commitment to physical mechanism and causal explanation, which blocks its ability to encompass both the noneffective reach of semantics and the normative nature of existential commitment.[16]

It is considerations of this sort that underwrite the first sentence in the book—that genuine intelligence, or anything worthy of the AGI label, will not be achieved with "more of the same": increased processing power, accelerated practical development, new technoscientific research of the sort that has been conducted to date. Contemporary research tends to focus on effective mechanisms, algorithms, architectural configuration, etc.—none of which are of the right conceptual *sort* to lead to insight about deliberation and judgment.

What *could* lead a system to judgment? Child-rearing is an instructive example. Developing judgment in people requires steady reflection, guidance, and deliberation over many years—interventions, explanations, and instruction in situations where judgment is required, even if that need is not superficially evident. What are those situations, where

16. Some will argue that psychological theories of memory are legitimately scientific and crucially traffic in the situations or phenomena that the memories are about (such as childhood experiences). The difficulty is that, although such accounts may help themselves to the contents of the memories, they cannot, within a mechanistic conception of science, explain or give an account of what makes those distal in-the-world situations be the memories' contents. (It would not do, of course, to say that those are the situations that the subjects describe when the memories are invoked or reported on; the problem just recurs—how can a mechanistic scientific account explain what their statements are about?)

guidance and instruction is warranted? How is it determined that (and what kinds of) intervention and guidance are required? These are questions with which every parent wrestles on an ongoing basis. The issues, symptoms, and recommended courses of action are deeply embedded in human culture; the patterns of behavior and thought that we hew to have been crafted over centuries of civilization (including in the religious traditions that have served as stewards for ultimate questions in all major historical civilizations). The stewards of a child's education—parents, schools, religious institutions, mentors, literatures, communities, and so on—bring to bear attention, wisdom, and reflection that build on this cultural legacy, skills distilled over many generations, and handed down from one generation to the next.

There are requirements on the child, as well. At each stage, a child must have some ability, increasing as they mature, to reflect on their own behavior, to "step outside" the situations they are engaged in enough to be able to assess the appropriateness and consequences of acting or responding in various ways, and so on. Notably, there is nothing in principle limited about what sorts of consideration might be applicable or worth considering in any particular case. There are maxims that are applicable; but any attempt to put them into articulated form would almost surely be vapid and contextually insensitive.

◆   ◆   ◆

What does all this mean in the case of AIs and computer systems generally? Perhaps at least this: that it is hard to see how synthetic systems could be trained in the ways of judgment except by being gradually, incrementally, and systematically enmeshed in normative practices that engage with the world and that involve thick engagement with teachers ("elders"), who can steadily develop and inculcate not just "moral sensibility" but also intellectual appreciation of intentional commitment to the world.

Is this a realistic route forward for AI? That is not for me to say. As I have made clear since the beginning, it is not my aim to bring the imagination of AI designers to a halt. But if synthetic judgment is an end toward which we want to aim,[17] these are not only the sorts of goals that will need to be embraced, but also the sorts of methods that may need to be undertaken. If that proves to be correct, attaining synthetic judgment will require a profound shift in the nature of research and a radical widening of approach. One of the aims of laying out the seven standards in chapter 8 is to suggest what would need to be addressed in order to make good on that original AI dream.

In passing, I should say that I expect advocates of evolutionary explanation and deep reinforcement learning to suggest that we could approach this process with appropriate "reward structures" and/or "threats," designed to mimic cultural and personal

---

17. Whether we do or not is a question on which I am not taking a stand here—though it is hard to imagine current efforts to develop AGI drawing to a close.

development. For at least two reasons, I doubt that such strategies will succeed. For one thing, they seem liable to drift into calculative reckoning about morality, rather than engagement with the world of the sort that will lead a system out of commitment to its internal representations and toward deferential commitment to the world. In addition, it seems unlikely that any simple exploitation, by a regimen of rewards, of even genuine existential vulnerability would suffice. Famously, you cannot "buy" generosity or kindness in people through any simple behaviorist pattern of stimulus and response, reward and punishment. There is no reason to suppose that the situation would be any different for a synthetic creature.

Judgment requires a finely-honed sensitivity to epistemic and ontological subtleties in ontological webs of practices and entities whose importance has been forged through cultural and historical processes beyond any one person's comprehension. I find it unlikely that judgment could be instilled or inculcated in a synthetic device through anything less than a moral and ethical upbringing and steady process of education, apprenticeship, enculturation, guidance—perhaps even friendship.

12 — Application

The impact of the seven points laid out in the last chapter can be appreciated by applying them to three issues, almost technical in nature, of critical importance.

12a · Reference

The first has to do with the nature of reference. I have said throughout that judgment requires holding registrations to account. This requires commitment to the world being registered. Commitment to the *registration* is insufficient, as I have emphasized throughout; so too, we can now see, is commitment to the world *as registered*. Rather, what is required is commitment to *that which is registered*. It is not just the difference between registration and registered that matters, that is—effectively a difference between sign and signified, between map and territory. Rather, the point is metaphysically more fundamental: there will always be a surplus to that which is registered, beyond its being as it is registered as being. You may be a person, but there is more to you than how my registration of you as a person takes you to be. And it is to that complete you that I refer, if I refer to you. To register an argument as brilliant, I must refer to it—whether or not it is brilliant, needless to say,[1] but also whether or not it is even an *argument*. In general,

---

1. If it is not brilliant, then I will have said or thought something false; but for that to be the case what I said must still be a statement, a thought, with the requisite stature to *be* false.

in order to fix on the world, reference and thought must "go on through" any registrational schemes used to parse or make sense of it, to reach the world itself. It is that patch of the world, not the world-as-registered, to which our commitment binds us, to which deference is owed. This is true—and this is the critical part, and the difficult one—even if the criteria used to individuate what we refer to have their roots in the registration practices employed.[2]

Figure 9[3]

As a possibly helpful analogy, consider the Mondrian painting shown in figure 9, and the descriptive term "the large grey rectangle toward the upper left."[4] Call this term α. Suppose, counterfactually but plausibly, that figure 10 is an accurate representation, at a very high level of magnification, of a

2. A realist who presumes that object ontology inheres in the world independent of human practice can view reference as attaching to a pre-existing object, and truth as having to do with whether or not that object exemplifies a property claimed of it. That option is unavailable to anyone with constructivist ontological sympathies. And the consequences of its not being available are profound. It is a challenge of the utmost importance to anyone attempting to make good on a metaphysical view with constructivist leanings to ensure that the commitment of judgment is to that-which-is-registered, not to the world-as-registered.

3. Piet Mondrian, *Composition with Red, Yellow, Blue, and Black,* (1921).

4. Red in the original.

portion of the top edge of that rectangle. At least arguably, greyish patch A falls within the referential scope of α, and dark patch B perhaps not (whether whitish patch C does or does not is less clear), even though A and B lie without and within, respectively, the boundary of the most plausible actual (idealized, Platonic) rectangle associable with α. Were Mondrian to instruct an assistant to "erase the rectangle," for example, it is unarguable that A should be removed.

The image is just an analogy. The point is that anything that is registered (a person, a gambit, a nation) will outstrip in richness not only what is grasped about it by its registrar, but also what is included within any idealizations used to identify it as the referent, and ultimately what is humanly graspable at all. *Only if that is true can registrations be held to account.* The point may sound mystical, but is in fact commonsense: that which we understand will always transcend our understanding of it—*including its "it-ness,"* its boundedness and identity as an object.

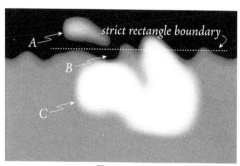

Figure 10

I said earlier that one of the things that differentiates a reckoning system from a system with judgment is that the former is content to operate on registrations of the world, whereas the latter must be committed to, must be concerned with, must care about the world thereby registered. What this last discussion emphasizes is that, in order to be accountably committed to that world, a registering system must "tread lightly"—holding its registrations and registration schemes in perpetual epistemic and normative abeyance, always prepared to let go of them in order to do justice to that which they are used to register. In this case a Buddhist phrasing is apt: you cannot reach the world until you "relinquish attachment" to the registrational machinery you employ to find it intelligible.

12b · Context

The second point has to do with an issue that has plagued AI research since the beginning—in both its first- and second-wave iterations—and that DARPA has identified as criterial for AGI, and imagines being part of "third-wave" AI: the notion of **context**.[5] The issue is not merely one of having a computer system use context-sensitive structures and symbols in appropriate ways, such as indexicals and perspectival descriptions (analogs of "today," or "the medium in this drive," etc.), but of configuring a system's ⌜deliberation⌝ to be appropriate to the wider situation at hand beyond what is immediately represented, either explicitly or implicitly—neither ignoring facts or phenomena that are in fact pertinent, nor wasting time and resources exploring any (or at least very many) of the infinite number of alternatives that "in theory" might be relevant, but in fact are irrelevant—and that anyone with a modicum of common sense would recognize as irrelevant.

We can approach the topic by considering the technical implications of part of the epistemological critique of first-wave AI: that thinking emerges from "a horizon of ineffable knowledge and sense-making." To be attuned to context is not simply being able to consider other factors from within a given registration scheme, or even being able to shift from

---

5. John Launchbury, "A DARPA Perspective on Artificial Intelligence," https://www.darpa.mil/attachments/AIFull.pdf

one registration scheme to another, as if stepping from one well-defined island to another, as in GOFAI's image of logical reasoning. Dealing skillfully with context—and this is really the point—is not a postregistrational, postontological skill. On the metaphor of chapter 3's figure 6 (p. 34), contextual awareness requires the ability to move continuously and ineffably in the submarine topology—stewarding the system's embodied and embedded participation in the world, registering only if, when, and as appropriate.

It has long been recognized as an unassailable fact about genuine intelligence that *anything* can be brought to bear if pertinent. There is no bound or specifiable limit to what might be crucial in any given circumstance, relevant in any particular line of reasoning. The present analysis ups the ante: there is no assurance that any registrational scheme will be adequate to an unexpected circumstance. It is conceptually impossible, in this English language, to describe a situation that cannot be registered, needless to say. As a gesture in that direction, however, we can imagine situations that put pressure on the idea of any preconceived registrational repertoire being adequate for training machine learning systems: a damaged aircraft crash-landing over one's head while driving, the ethics of murder in a society where people regularly back up their brains online, the nature of a universe with two spatial and two temporal dimensions rather than three and one, and so on.

The only "ultimate context" is the world as a whole—something that no registration scheme will ever encompass. Sensitivity to context, in other words, requires far more than choosing from a prefigured set of registration schemes. *Context-sensitivity is not merely a question of having a world model*, in other words. No such model can be adequate to all potential circumstances. Rather, sensitivity to context requires being able to choose or develop a model (registration scheme) adequate to whatever circumstances are at hand, backed by resolute commitment to hold that model (scheme) accountable to the world it is a model of.

The world requires commitment to keep in view. Contextual awareness must be based on such commitment. As such, *conceptual sensitivity requires judgment. It will never be achieved by mere reckoning.*

12c · TW

Finally, it is instructive to take stock of where this long analysis has taken us.

I claimed in chapter 3 that the ontological critique of GOFAI cut the deepest.[6] It was not a critique *of ontology*, needless to say, but of AI systems, having to do with the ontological assumptions they made about the worlds they were tasked with understanding. In framing it, I used the term "ontological" to mean roughly "what there is"—that is, the furniture of the world, what objects, properties, and relations exist.

I also said, in the introduction, that one of the most important contributions of second-wave AI is the fact that it gives us a window onto a different ontological view. But from where we have reached we can see that that was an in-

---

6. The phrase "ontological critique" is from Dreyfus's *What Computers Can't Do* (1972); he used the term for a critique of AI similar to mine, although, somewhat surprisingly, he framed his version rather epistemologically—as having to do with information, data about the world, understanding, etc., rather than with the structure of the world per se ("the *data* ... must be discrete, explicit, and determinate" [118], "everything essential to intelligent behavior must in principle be *understandable* in terms of a set of determinate independent elements" [118], "the world can be *analyzed* as a set of facts—items of *information*" [130], etc.; emphases added).

adequate characterization. The impact of ML is far greater. These new systems, and the experiences we are gleaning through building them, are teaching us *about ontology itself*—about what objects are, about how individuation arises, about what the world is like "underneath" the objects (i.e., in more detail than is grasped in the abstractions in terms of which the objects, properties, and relations are individuated). Second-wave AI, in other words, is providing us with insight into the very metaphysical foundations of ontology.[7] I said at the outset that the book would have a strongly ontological flavor. Had we had the requisite distinctions in front of us at that point, I could have said: a strongly ontological *and metaphysical* flavor.

Second-wave AI, we have seen, is demonstrating something I have said for many years: that ontology emerges in the context of registration practices; it is not the prior, pregiven structure of the world.[8] Moreover, and directly

---

7. Nonstandardly, I use the terms "metaphysics" and "ontology" distinctly: "ontology" to refer to what there is—to the furniture of the world, to reality *as we register it*; and "metaphysics" to refer to (and, when appropriate, to the study of) that which undergirds ontology—the world, the foundations and fundaments of objects and ontology.
8. This is not to say that ontological items or entities—objects, properties, states of affairs, etc.—are *ways of taking the world*. Tables, detente, machinists, and love affairs are absolutely not merely epistemic entities—not thoughts, not representations, not ideas, not ingredients in our registration schemes, not patterns of perception, not inferential processes, not types of data structure, not any other purely cognitive or epistemic or machinic configuration. In that sense the view I am defending is stubbornly realist: all these entities are things beyond us, things in the world. The point, though, is that the individuations and abstractions and schemes in terms of which this furniture of the world is framed (individuated, categorized, discretely conceptualized, ontologically "parsed") are the result of social and individual processes of registration. Tables, détente, people, etc., are not "ways of taking the world." They are *the world taken a certain way.*

applicable to AI, registration, including determining an appropriate registration scheme, is among the most important characteristics—if not *the* most important characteristic—of intelligence itself. To put it bluntly: it is not sufficient for AI, and the broader study of intelligence of which it is a part, merely to *make assumptions* about ontology. It must *explain* ontology.

Investigating registration, and thus moving the subject matter of ontology into the research agenda of artificial intelligence, are among the most important philosophical and scientific consequences of recent developments.

This much we have learned from ML and second-wave AI:

1. We need to revise our understanding of concepts, and of conceptual reasoning, to appreciate the power that concepts derive from being drenched in non-conceptual (submarine) detail.
2. We need to understand registration, and recognize registering the world appropriately as possibly the most important characteristic of intelligence.
3. It is insufficient for AI, and the broader study of intelligence of which it is a part, to assume that intelligence is a capacity of systems deployed in an ontologically structured world. Ontology is an achievement of intelligence, not a presupposition.

But there is more. What ups the ante for AI is that genuine intelligence—intelligence at the highest level, anything of the sort that could underwrite judgment—requires *understanding* these ontological points, in at least a tacit sense—enough for them to both normatively and pragmatically govern its behavior. As I have emphasized throughout, judgment requires commitment to what we might call a third level of representation: not to the first level of the representation or registration itself (not, as I have said since the beginning, to a data structure, or to an image of a person, or to a description or term); and not even to the second level: the world *as registered* (as a table, machinist, etc.).[9] I am not convinced that either first- or second-wave AI systems have yet made the transition from the first of these to the second. But anything approaching judgment requires a third, much more demanding version: commitment to *that which is registered* a certain way (as a table, détente, or whatever). As always, "world" in this statement does not refer to what we ordinarily think of the world as being—a planet circling the sun, a profusion of social and political arrangements, a biological ferment. Rather, the world, in the only sense that makes sense for this third level of commitment, is that which underlies and underwrites all these things—the ground of being, that which is One.

I know that describing the world in terms of the "One" or the "ground of being" may sound mysterious, if not outright mystical. But the point is not difficult to understand, and neither problematic nor spooky. An easier way to say it, which I have used in this book many times already, is to say that the third level of commitment must be to *that which* we register—to *that which* we take to be a table, in the case of a table; to *that which* we register as a person, in

---

9. An adequate philosophical analysis of these things would require consideration of traditional categories of content, sense, reference, etc. But the distinction being developed here—between a patch of the world registered as a chair and that which is registered as a chair—does not normally arise, because the referent of the singular term "the chair" is taken to be the object that is the chair, on an assumption that the object exists and has identity independent of its exemplification of the chair property—an assumption not available in the metaphysical view I am defending.

the case of a person; and so on.[10] Not (first stage) to the representation "table" or "person," and not (second stage) to *the table as registered as a table* or to *the person as registered as a person* (since that would not leave us in a position to be able to say "that isn't a person after all"—or even "what you are taking to be a table isn't even a coherent object"). Rather, only if we are committed (third stage) to *that which* we take to be a table, person, and so on, will we be in a position to hold the registration to account.

In my teaching, I introduce the acronym "TW," explaining it first as shorthand for "the world" in the sense that I am using it here—the "ground of being," the One, and the like. But as the conversation progresses it proves helpful to shift the connotation of the acronym (if not its meaning), and say that "TW" instead stands for "that which"—the "that which" toward which commitment, registration, etc., must be oriented.

Deference must be deference to *that which* we register, inhabit, care about, live in, and so on. Registration must be of *that which* we take to be a table, a person, or anything else. Holding registrations and registration schemes accountable to the *that which* is what it is to hold them accountable at all.

---

10. "That which" refers to that which plays the role of bare particulars in philosophical substance theories, without projecting object identity into the world as an ontological given.

13 — Conclusion

Using ML and other computational techniques, including many from first-wave AI, we are building AI systems of unsurpassed reckoning power. The juggernaut will only accelerate. In many realms these systems will outstrip us in reckoning prowess, if they do not do so already. For good or ill—usually for good, we can hope, but often for ill, we can also be sure—we will increasingly delegate tasks and projects to them. Mining vast troves of data, and employing computational power beyond imagining, they will increasingly dominate the substrate and infrastructure of life on the planet.

What I do not see, however, is anything on the horizon—in scientific or technological or even intellectual imagination—that suggests that we are about to construct, or indeed have any ideas as to how to construct, or are even thinking about constructing, systems capable of full-scale judgment:

1. Systems existentially committed to the world they register, represent, and think about
2. Systems that will go to bat for the truth, reject what is false, balk at what is impossible—and know the difference
3. Systems not only *in* and *of* the world, but *for which there is a world*—a world that *worlds*, in the sense of constituting *that to which all is ultimately accountable*

4. Systems that know that the world that hosts them, the entities they reason about, and all of humanity and community as well, must be treated with deference, humility, and compassion

It is this kind of judgment, I believe—a seamless integration of passion, dispassion, and compassion—that ultimately underwrites what matters, not just about the human, but about the sacred, the beautiful, and the humane.

It is this kind of judgment, I am arguing, that must be the aim of any project wanting to construct "artificial general intelligence." I do not believe it is inherently beyond the reach of synthetic construction. But it is also not an incremental advance beyond first- or second-wave AI—beyond the systems we have devised to date. There are profound differences between judgment and reckoning—especially the sorts of reckoning we at present have any capacity to construct. While predictions are a fool's errand, I cannot see our synthesizing full scale judgment, if indeed we are ever capable of doing so, in anything that anyone could call the short term. Even minimal progress in that direction will require strategies wholly unlike any that have been pursued in first- and second-wave AI.

But we are learning. Second-wave AI has brought to our attention the inadequacy of the formal ontological assumptions that underlay GOFAI. From its successes, and building on the insights of a wide diversity of other fields, we should have renewed respect for three interlocking and inexorable facts:

1. The world is surpassingly rich—far transcending any ability to capture it in formal symbols, or for that matter in any discretely conceptualized structure.
2. All (especially conceptual) registration is inevitably skewed, partial, and interest-relative.
3. By exploiting yet also transcending the limitations of registration, genuine intelligence is committed to and directed toward the one and only world.

The combination of the three implies that any system that steps from one registration scheme to another, or that deals with (or is used in) different circumstances, or that attempts to integrate information gleaned from different projects, must ground its deliberations in full-scale judgment *at every step of the inferential chain*, in order to ensure that its representations never take leave of accountability to the world. These are strong but sobering conclusions—but they are straightforward consequences of how the world is.

Where does that leave us? We should be humbled by GOFAI's inadequacy, given the depth of the insights on which it was based. We should be cautious about the successes of second-wave AI, mindful of its limitations and restrictions. But mostly we should stand in awe of the capacity of the human mind, and of the achievements of human culture, in having developed registrational strategies, governing norms, ontological commitments, and epistemic practices that allow us to comprehend and go to bat for the world as world.

References

Adams, Zed and Jacob Browning, eds. *Giving a Damn: Essays in Dialogue with John Haugeland*. Cambridge, MA: MIT Press, 2016. [108]

Athalye, Anish et al. "Synthesizing Robust Adversarial Examples," *Proceedings of the 35th International Conference on Machine Learning*, Stockholm, Sweden, PMLR 80 (2018). [57]

Brooks, Rodney. "Intelligence Without Reason," MIT Artificial Intelligence Laboratory Memo 1293 (1991). [14]

Dennett, Daniel. *The Intentional Stance*. Cambridge, MA: MIT Press, 1987. [62]

Doyle, Jon. "A Truth Maintenance System," *Artificial Intelligence* 12, no. 3 (1979). [51]

Dretske, Fred. *Knowledge and the Flow of Information*. Cambridge, MA: MIT Press, 1981. [35, 99]

Dreyfus, Hubert. *What Computers Can't Do: A Critique of Artificial Reason*. New York: Harper & Row, 1972. [xix, 23, 140]

Ekbia, Hamid. *Artificial Dreams*. Cambridge: Cambridge University Press, 2008. [37]

Evans, Gareth. *Varieties of Reference*. Oxford: Oxford University Press, 1982. [29, 30]

Feldman, Jerome and Dana Ballard. "Connectionist Models and their Properties," *Cognitive Science* 6, no. 3 (1982). [23]

Fodor, Jerry. "Connectionism and the Problem of Systematicity (Continued): Why Smolensky's Solution Still Doesn't Work," *Cognition* 62, no. 1 (1997). [72]

Friedman, Thomas. "From Hands to Heads to Hearts," *New York Times*, Jan. 4, 2017. [123]

Gärdenfors, Peter, ed. *Belief Revision*. Cambridge: Cambridge University Press, 2003. [51]

Haugeland, John. "Analog and Analog," *Philosophical Topics* 12, no. 1 (1981). [31, 74]

———. *Artificial Intelligence: The Very Idea.* Cambridge, MA: MIT Press, 1985. [7]

———. ed. *Mind Design II: Philosophy, Psychology, Artificial Intelligence.* Cambridge, MA: MIT Press, A Bradford Book, 1997. [14]

———. "Truth and Rule-Following," in *Having Thought.* Cambridge, MA: Harvard University Press, 1998. [88, 94]

———. "Truth and Finitude," in *Dasein Disclosed.* Cambridge, MA: Harvard University Press, 2013. [94]

Hutto, Daniel. "Knowing What? Radical Versus Conservative Enactivism," *Phenomenology and the Cognitive Sciences* 4, no. 4 (2005). [27]

Launchbury, John. "A DARPA Perspective on Artificial Intelligence," https://www.darpa.mil/attachments/AIFull.pdf [138]

LeCun, Yann, Yoshua Bengio, and Geoffrey Hinton. "Deep Learning," *Nature* 521, no. 7553 (2015). [47]

Levesque, Hector. *Common Sense, the Turing Test, and the Quest for Real AI: Reflections on Natural and Artificial Intelligence.* Cambridge, MA: MIT Press, 2017. [75]

Lighthill, James. "Artificial Intelligence: A General Survey," in *Artificial Intelligence: A Paper Symposium*, Science Research Council, 1973. [53]

MacIntyre, Alasdair. *After Virtue.* Notre Dame, IN: Notre Dame University Press, 1981. [111]

———. *Whose Justice? Which Rationality?* London: Duckworth, 1988. [118]

Marcus, Gary. *The Algebraic Mind: Integrating Connectionism and Cognitive Science.* Cambridge, MA: MIT Press, 2001. [75]

Maturana, Humberto and Francisco Varela. *Autopoiesis and Cognition: The Realization of the Living.* Dordrecht: Reidel, 1980. [27]

McCulloch, Warren. "What is a Number, that a Man May Know It, and a Man, that He May Know a Number?," *General Semantics Bulletin*, no. 26/27 (1960). [3]

McDowell, John. *Mind and World*. Cambridge, MA: Harvard University Press, 1996. [29, 93]

Millikan, Ruth. "A Common Structure for Concepts of Individuals, Stuffs, and Real Kinds: More Mama, More Milk, and More Mouse," *Behavioral and Brain Sciences* 21, no. 1 (1998). [99]

————. "Pushmi-pullyu Representations," *Philosophical Perspectives* 9 (1995). [100]

Pater, Joe. "Generative Linguistics and Neural Networks at 60: Foundation, Friction, and Fusion," plus comment articles, *Language* 95. no 1 (2019). [75]

Piccinini, Gualtiero. *Physical Computation: A Mechanistic Account*. Oxford: Oxford University Press, 2015. [10]

Rosch, Eleanor, Francisco Varela, and Evan Thompson. *The Embodied Mind*. Cambridge, MA: MIT Press, 1991. [27]

Searle, John. *Speech Acts: An Essay in the Philosophy of Language*. Cambridge: Cambridge University Press, 1969. [86]

Smith, Brian Cantwell. "The Owl and the Electric Encyclopaedia," *Artificial Intelligence* 47 (1991). [37]

————. *On the Origin of Objects*. Cambridge, MA: MIT Press, 1996. [xvi, 8, 18, 26, 29, 32, 35, 41, 67, 81, 98]

————. *Computational Reflections*. Forthcoming. [4, 9, 128]

————. "Rehabilitating Representation," unpublished. [15, 35]

————. "Solving the Halting Problem, and Other Skullduggery in the Foundations of Computing," [12]

————. "The Nonconceptual World," unpublished. [64, 68]

————. "Who's on Third? The Physical Bases of Consciousness", unpublished. [121]

Strawson, P. F. *Individuals*. London: Methuen, 1959. [15, 99]

Suchman, Lucy. *Human-Machine Reconfigurations: Plans and Situated Actions*. Cambridge: Cambridge Univ. Press, 2007. [27]

Sutton, Rich. "The Bitter Lesson," http://www.incompleteideas.net/IncIdeas/BitterLesson.html (retrieved March 16, 2019). [63]

Thompson, Charis. *Getting Ahead: Minds, Bodies, and Emotion in an Age of Automation and Selection*. Forthcoming. [122]

Thompson, Evan and Francisco Varela. "Radical Embodiment: Neural Dynamics and Consciousness," Trends in *Cognitive Sciences* 5, no. 10 (2001). [27]

Weizenbaum, Joseph. *Computer Power and Human Reason: From Judgment to Calculation*. New York: W. H. Freeman and Company, 1976.

Winograd, Terry and Fernando Flores. *Understanding Computers and Cognition: A New Foundation for Design*. Norwood, MA: Ablex Publishing, 1986.

Index

The letter *n* following a page number denotes a footnote; *f*, a figure; *s*, a sidebar.